The Wildlife Pest Control Handbook

By

Philip J. Nichols

PERKIOMEN VALLEY PRESS
Part Number 350004

First Edition
First Printing March, 1993

Library of Congress Catalog Card Number: 93-92554
ISBN: 0-9635911-0-X

I dedicate this book to fellow trapper Dennis Byrd

Contents

Contents

Chapter Zero

Ramblings and Such

Why are trappers, wildlife biologists and other outdoor types launching commercial ventures that focus on servicing wildlife problems? Increasing numbers of outdoor-oriented people having the skills and enthusiasm to do the work, are recognizing the proverbial "ground level opportunity".

I had never quite figured out what to do with my life until I became a Wildlife Pest Control Operator (WPCO). It is an enjoyable job for anyone with an interest in wildlife. Most WPCO's are part timers, but more are finding enough work to go into the business full time. A few aggressive entrepreneurs already have six figure incomes. All the signs point to a healthy and growing service industry.

Why all of a sudden is there so much work for WPCO's? Wildlife populations in cities and suburbs get out of control because they are not regulated. Traditional controls (hunting, trapping and predation) have been replaced by road kills and disease. Government agencies used to provide pest control services, but now the workload is way beyond any hope of control through taxpayer dollars. The private sector is taking up the slack.

Natural wildlife populations are largest in spring and summer, diminish during autumn, and finally reach the lowest point in winter. The conditions in our cities and suburbs promote unbalanced wildlife populations due to the abundance of food and shelter, the absence of population controls, and the overall protection available. This situation creates jobs for WPCO's with the best opportunities being where human populations are greatest.

Business and property owners are usually responsible for getting wildlife problems taken care of. Their immediate need is to find a cost effective WPCO who can solve their immediate problem, and design a strategy to prevent future problems. Finding these customers is essential for a successful WPC business.

It is becoming easier for the customer to find a WCPO due to increased publicity in the media (newspaper and TV), and more effective marketing by WPCO's. On the side of the WPCO, there are trade associations, books, and magazines that have surfaced in the last five years. These resources provide information about new market opportunities.

How does the WPCO deal with a typical situation such as a large and ornery raccoon in the chimney? The professional WPCO removes the animal, its nest, feces and any other waste, disinfects and thoroughly cleans the area, caps the chimney, and does whatever work is necessary to prevent future problems. The animal is treated humanely, the customer is relieved, and the WPCO is compensated. All WPC work is a combination of observation, knowledge of the animal, diagnosis of the problem, followed by effective action.

Luxury items such as fur, take a beating in a depressed economy. Fewer fur trappers results in more furbearing animals. More animals cause more problems, resulting in more business. I believe the WPC market will be just as strong even after the fur market recovers because furbearers are not harvested where the problems are most common, in cities and suburbs. Fur has always been a speculative commodity, with pelt values going up and down over time. The animal rights advocates believe that their activism caused the depressed fur market. This may be partially true, and conversely implies they are instrumental in causing the WPC boom. Man has been a hunter-gatherer for tens of thousands of years, and this spirit will always be a part of our being.

This Wildlife business is not for the squeamish, stupid, lazy or arrogant. It is at times dirty and dangerous, requires high ethical standards, along with ingenuity, keen powers of observation, physical stamina, and perseverance to succeed and stay on top. These are familiar traits to trappers.

Some basic business skills are also needed, involving paperwork, record keeping and other such foolishness. It is not all that complicated. Put simply, all transactions get recorded (all money spent and all money received). If the made is more than the spent, you are headed in the right direction. More details will follow in the business section.

The successful WPCO is professional in every sense of the word. Some trappers will have to make adjustments. The generic image of the trapper is that of a resourceful loner, maybe taking too many risks. By contrast, the outgoing businessman works in groups, and gambles less. It would be ideal to combine the best of both. Compare yourself to someone you consider a professional, and it doesn't have to be a doctor or lawyer. Work on developing professionalism. Here are some potential areas to consider: time management, vocabulary, writing skills, appearance, discipline (work, eating, drinking, etc.) attitude and study of wildlife literature.

In the past, trappers parlayed with land owners, game wardens, fur buyers and fellow trappers. WPCO's have one more to bargain with, that is more difficult and hard to figure out than all of the aforementioned combined : the customer! Understanding customers takes patience. Listen to what the customer says. The critter problem may really be a problem with the neighbor, or a lazy son. Don't believe the old saying "the customer is always right". After all, they call a WPCO due to their own inability to solve the problem.

Unfortunately, there is no internal alarm that goes off in the brain when a critter problem is actually some kind of people problem. Most customers are decent, but you will eventually have to deal with a jerk or two. Use the occasion to practice being patient and calm, even though they might benefit more from a good strangling.

The customer expects effective service, and you will be accepted as a professional so long as their expectations are met. However, what is the reaction to "Cooter" shuffling up to the door, mouth full of chew, barely speaking in syllables, unshaven and reeking of skunk? If the problem gets taken care of, the customer may be grateful that both the critter and "Cooter" are gone. I want every chance to provide more service, or get referrals to others having similar problems. Cooter will miss out on these opportunities.

Think about the kind of market there is for WPC service where you live. Some areas are mother lodes of opportunity, with plenty of critters causing endless problems, and no competition. If there are enough customers willing to pay for your service, and they know how to find you, it will not be difficult to have a profitable business. Maybe your area does not have a market large enough to support a full time business. Do you move? Or is there a way to prospect and find or create some business without moving? How much effort, time and money can you afford while prospecting? How do you handle rejection, failure and hard times? Will you be pushed to even greater effort, or just say to hell with it?

One asset trappers have is the tempering we get from our encounters with nature: ice to slip on, mud to wear, swift cold water to fill our boots with, and a thick cloud of skunk musk for effect. We pity "normal" people who will never experience this peculiar combination of physical misery and happiness.

One final thought. If the quiet life is not for you, here is an opportunity to be an advocate. Trappers regard the world different from our "correct and civilized" neighbors. Lately, it has been our turn in the barrel, so to speak. Animal rights groups portray us as barbarian mutilators of helpless, endangered, cute little forest creatures (who talk to one another in high squeaky voices). These enlightened souls have evolved to the "next plane", while us subhumans are left to grovel in the dirt, preoccupied in mindless frenzy of torture and death. This stereotype is partly due to our own stubborn resistance to change, in a world where change has become the only constant. Slowly, the trapping fraternity is beginning to respond. The present economic reality of the fur industry has driven out most of the opportunists, dumb clucks and dim bulbs. Those remaining seem dedicated to the long haul. It will be a effort to educate both ourselves, and those who need to understand us. You have to speak up to be heard.

This book contains material from my experiences and represents the variety of work done where I live. The opportunities for other locations may be different, but the principles will be the same. Pay attention to details, observe, be creative, work hard and keep at it. There is a lot of material to absorb at once. Slow down and think about the things that are meaningful in the first reading. Later, go back and read the passage again. And, please don't cuss me for writing stuff you may already know. If it really bothers you, write your own damn book.

To Succeed In This Business Do You Have?

- Physical energy
- Mental toughness
- Spiritual tenacity
- Curiosity
- A sense of fun
- The will to improve yourself, and in turn, your fraternity

- A healthy sense of fear
- Personal knowledge of pain
- Self discipline
- Imagination
- The ability to manage time
- The ability to organize
- An eager obsession to learn everything possible
- The ability to know when to get help
- Lots of excuses

Chapter One

Business Considerations

There is a connection between enjoyment and success. It is obvious why trappers gravitate to a business where they are respected, and their trapping skills have significant value. The downside is having to learn the non-trapping skills a business owner must have. But it beats the hell out of having to work for someone else in a job you hate. Try to enjoy what has to be learned in the frame of mind that nothing will stand in your way to success.

We all have different definitions of success. Maybe you would be happy with plate of beans now and then, and a bone for old Blue. In this case, you can do just about anything you want, and you will by your definition, be a success. If your ambition is a bit greater, you will need to learn how to manage a business. Learning is not too complicated or expensive, but it does require regular, purposeful work.

A business is an "organization" which suggests something structured, neat, tidy, uncluttered and orderly. What if you are not an orderly person? Your desk is piled high with papers and all sorts of junk, but you have a steady stream of customers, which could mean you are doing something right. A disorganized desk is something everyone has to deal with. A sound structure makes it easier to "manage" your problems, little or big.

What is "business structure"? Look at the figure for a sample. No two are alike, and yours will be the product of who you are, what you do, and a host of other factors such as location, financial condition, personality, energy, luck and ambition. Anyone in business has a structure whether it's designed or not. You mind as well plan your structure since it will happen anyway.

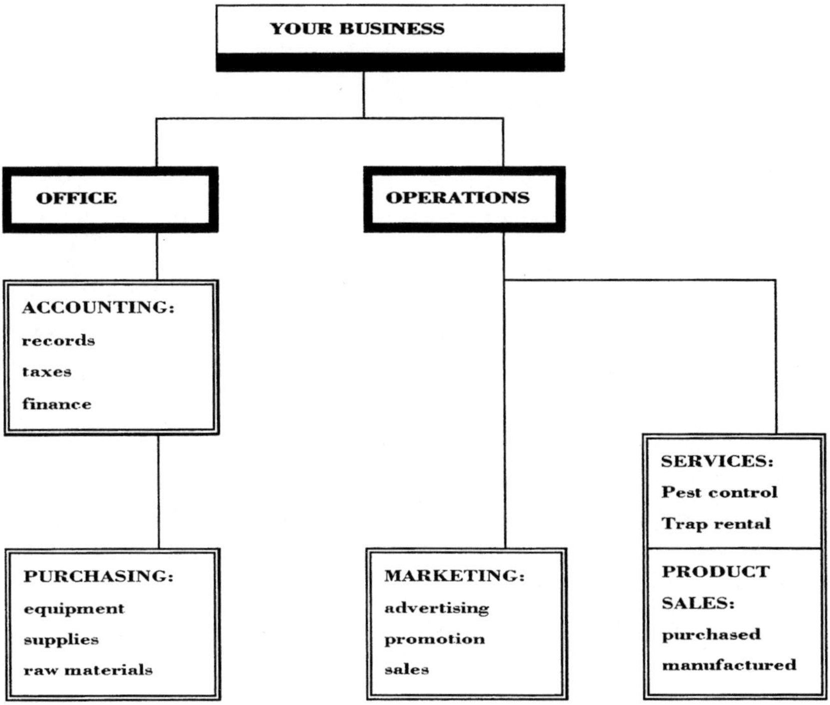

Sample Organization Chart

Keeping Records - Computerization

Records can be kept in your head, in filing cabinets, in a computer, or some combination of the three. I recommend keeping records in a personal computer because PC's are inexpensive, and learning to use one is not difficult. A small business should have programs for accounting, bookkeeping and word processing. In addition, the WPC can use a program to track jobs, especially if there are employees doing those jobs. You will want to know the results of various sets, baits, lures, etc. The program can provide timely, accurate information in report form. Reports can show where mistakes are repeated, ways to improve operations, and any other information you need.

If you can, invest in an IBM compatible computer. Don't be spooked by computer language. Computer talk may seem mysterious, but is no worse than trappers talk. In any case, concentrate on learning how to use it.

How long does it take to learn how to use a computer? How much is your time worth? The computer can be a working tool in a very short time. Spend one hour a day with it and you will be amazed at how soon it becomes useful. I know, you are saying; work, family, hobbies, sleep, there is a limit to what can be squeezed into 24 hours. Look for time wasters. For example, most of the TV watched is time wasted (if you don't have to look in the TV guide to know what is on, you are watching too much). Be honest with yourself when defining the activities in your daily routine that are not productive. Focus your energy and time on making the goals you set.

How much money should you spend on a computer system? Shop around and talk to people. For $500 you can get a used older system (IBM-XT and compatibles) that, while slow by todays standards, will handle all of the business functions. A modern high speed system (IBM-AT 386 or 486 and compatibles) costs about $2000 new and has these components (not including the extras):

1. HARDWARE

   A. Personal Computer: 386SX-33MHz

B. 2MB RAM
C. 80MB Hard Drive
D. 1.2 MB Floppy Drive (uses 5 .25 "disk)
E. 1.44 MB Floppy Drive (uses 3 .5 " disk)
F. Printer port and printer (24 pin dot matrix)
G. Mouse port and mouse
H. VGA color monitor

2. SOFTWARE

A. DOS (operating system)
B. Word Perfect (word processor)
C. Quick Books (accounting for small business)

3. EXTRAS

A. Modem (to talk with other computers over telephone line)
B. Trapping Software (see appendix for source)

This PC system has room to add all kinds of hardware and software enhancements. In the future, WPCO's will hook up their computers' modem to the phone line to get wildlife information and to network with each other, exchanging ideas on problem solving, sources of supply, wildlife research and who knows what else? This is a new frontier for those who see the opportunity and are progressive enough to take advantage of it. "Opportunity is never lost . If you don't take it, don't worry, someone else will" (anonymous quote).

Computer literacy will be essential if you want to compete in the years to come. And if you tell yourself that you do not have the (fill in blank with brains, money, time, etc.), congratulations. You have just programmed yourself to fail.

More Basics

If money is scarce (har har har), buy only the essentials. You need a business checking account, a dedicated telephone, some way to answer the telephone when you are not there, and some files for records such as accounts payable (who you owe money to) and accounts receivable (who owes you money). You don't need new, costly filing cabinets to start out. I use old cardboard boxes to keep files in. I have files for information about wildlife, newspaper cutouts, articles from trapping newspapers and magazines, animal rights, wildlife disease, and around thirty related topics. You also need a place to work, with enough room to store all the "stuff" you have, along with space to grow (you will be getting more "stuff").

Your business needs a name. You can have some fun picking a name, just remember you have to live with the choice. I don't have a clue as to how you can come up with a really great name, but I can tell you to clear it with the state first. My state has a "fictitious names registry" that keeps track of names so there are hopefully no duplicates.

Financial Concerns

The business checkbook is the key to managing finances. Fill out the stubs exactly the same every time. If you do not keep an accurate and up to date account of your money, maybe you are not cut out to be in business and would be better off working for someone else.

1504	BAL FWD		
DATE	DEPOSITS	120	.00
Jan 14,1992 680.00			
TO ABC Real Estate			
	TOTAL	800	.00
FOR Jan. Rent	THIS CHECK	450	.00
	OTHER		
	BALANCE	350	.00

Business Checkbook Stub Sample

Take a look at what is filled in on the stub: commit yourself to the positive habit of filling in the same information for every transaction, at the time the transaction is made. Don't wait and don't make excuses. Just do it, and use legible hand writing. I have wasted a lot of time checking for arithmetic mistakes that turned out to be careless handwriting.

Reconcile the checkbook as soon as the bank statement arrives, and this miserable but necessary chore will be easier when the business grows and things get more complicated. Then is no time to play catch up. The checkbook system comes with instructions on how to reconcile the statement. It is not difficult, but it does take your concentration for awhile. Things to look for if the numbers do not match are checks that you have payed out, but have not been cashed for over a month. Also look for bank charges, entries made twice or not at all, and mistakes. The best solution for accounting miseries is to get an accountant to explain things. Better yet, get one to do it for you. No matter who does the accounting, the owner (you) must be able to look at the books and know what is going on.

When you apply for a business checking account, the bank will ask what kind of legal business structure you have. The three possibilities are: sole owner, partnership, and corporation. The least complicated and least expensive option is sole ownership. It helps if you have an existing relationship with the bank, but just remember that you are the customer, and the bank wants to sell it's services to you. It makes sense to shop for a bank where you are best served. You will want to ask how much the checking account costs, how to get a commercial line of credit, overdraft protection and questions about any other bank service you might need. Make sure you have both your personal name and the business name on the signature card. Don't be shy when dealing with banks - they need your business.

If you are thinking about incorporation, get as much advice as you can. You will need a lawyer to do it, which costs between $400 and $1000 depending on the lawyer, and the state you incorporate in. There are too many variables to list here, but incorporation may be the thing to do if other people will be investing money in your business. I do not encourage outside investment until you have experienced losing your own money, but this is a personal choice.

The least expensive business entity to set up is the sole proprietorship. The difference between it and a corporation is the corporation is owned by stockholders, while the proprietorship is owned by you. The cost to set up a proprietorship is simply the cost of getting your name registered, along with any other local licenses or fees. The partnership is set up the same as a sole proprietorship. They differ in that you will have a partner. If you have no business experience, I do not recommend starting with a partnership.

A variation of the three business configurations discussed so far is the franchise, which helps small business owners get a sound start. Federal franchise laws require that all franchisors must give you a full and complete description of the business, training programs, services provided, number of franchisees, financial statements and audits. They must fully describe what will be required of you. It would be in your best interest to retain a lawyer before signing any franchise documents. There is one WPC franchise called "Critter Control" at 313-453-6300 that claims to have sixty offices throughout the US. If this kind of setup appeals to you, give them a call for more information.

Insurance for WPCO's

Most people cringe at the mention of "insurance". However, if are in a business where the service provided could be hazardous to the customers life, limb or property, a comprehensive liability insurance policy will protect you from financial loss, up to the limit of the policy. A basic policy will cost between five hundred and one thousand dollars a year, and will provide from one hundred thousand up to millions of dollars worth of protection, with some kind of deductible. You have to decide if it is worth it. Some jobs require insurance, such as providing a contract service to a municipality.

Think about the possibilities for damage to the customer and their property. Suppose you are up on the roof to remove a raccoon from the chimney, and your ladder falls, taking down the telephone wire, some shingles, and breaks a big picture window. Forget that you feel like an idiot, guess who is liable for the damages? These hazards are manageable, and I will tell you how to minimize risks later on. I bring them up here to impress you with the purpose liability insurance has in your business. Imagine this scenario: a customer calls and says there is a raccoon in his house, would you come and get it? You say OK, arrive at the scene, and find a very aggressive animal. Before you can say hoot, the animal attacks the customer and munches on his leg. You quickly capture the raccoon, which must be tested for rabies. The owner got bit because of his own stupidity, but he says it's your fault, and will sue you for pain and suffering. Does this sound far fetched? With liability insurance, you will have someone on your side to sort things out.

Conventional insurance companies do not have liability insurance for Wildlife Pest Control operators. This falls in the category of non-standard insurance. Some insurers will charge a minimum of one thousand dollars for this kind of policy, if they offer one at all. Ask your insurance agent for advice. If you don't get anywhere, call the Wyomissing Group at 1-800-222-3994 and request an Urban Wildlife Management Liability Insurance application. This company has experience serving WPCO's, and has tailored a policy to our needs.

Legal Matters

Permits, licenses, fees, etc. are a necessary part of doing business. My state (Pennsylvania) requires a Wildlife Pest Control permit, renewable annually for fifty dollars. In addition, Pennsylvania requires WPC operators to collect sales tax, which requires a sales tax license. Check for local laws and codes on file at the town hall. Zoning ordinance regulate where a business can be located, and what kind of business can be conducted. You can operate part time from your home if the business has a low profile, but expect neighbors to complain if there is noise, traffic (bunches of obnoxious people), odor (skunk?) or other unusual disturbances.

Whenever you take on a job, be sure that you are dealing with the person or group that is responsible (through ownership or other legal means) for the property. My permit stipulates that I can only do business with property owners. This policy protects both parties in the transaction.

If you must hire a lawyer, expect to pay $75 to $150 per hour for business related services. Try to get a fixed fee rate for a specific task. Better yet, do the task yourself. Many business related legal processes do not require a lawyer, just a little bit of research on your part. For example, you do not need a lawyer to get a copyright. A friend paid a lawyer $150 to get a copyright, not knowing that copyright is automatic when you create the work. Only if you register it with the copyright office in Washington do you pay a $10 fee. You simply fill out a form provided by the copyright office, and send them a sample of the work.

There is much free information available from the Small Business Administration in your state. You may be surprised. Retired business executives are available to help new business owners through the SCORE program. The SBA office nearest you can provide information about the SCORE program.

Pricing

How much to charge for your services depends on location, customers, and competition. My advertisements state "free estimates". I try to figure out how much time the basic job will take over the phone, and make sure the customer understands that an estimate is just that, and that extra service is extra expense.

The initial quoted cost to homeowners is relatively low, to prevent "sticker shock". Customers are especially receptive to my quote if they have been shopping around at conventional pest control companies, those specializing in bug and termite work and such. These places usually quote wildlife jobs on the high side. I want to convince the customer to invest in some exclusion work once they understand the benefits, but my initial goal is to close the sale. For example, I quoted $50 over the phone to remove a squirrel stuck in a chimney. After removing the animal, the customer agreed that it would make sense to put on a chimney cap for an additional $35. Total time on site was one hour. Remember, the initial quote of $50 was accepted without hesitation. Sometimes I will offer to clean a chimney since I am already up there, which is $50 more. These prices are in line for my area, customers, time on the job and competition. I know some part timers around here that can charge less because they do not have my overhead (insurance, etc.) expenses. It takes overhead to provide a professional full time service. A professional can calm all of the customers fears, answer all questions, address all concerns, and solve the problem at hand for a reasonable fee. Keep an eye on the competition, but don't lose any sleep. Customers that base their buying decisions on price alone usually turn out to be a pain in the rear anyway.

Purchased material that is used to solve a problem must be figured in the estimate. Anything that is purchased to do the job must be included in the bill. I charge a bare minimum mark up of at least 40% ($1.00 bought = $1.40 sold) I do not itemize on the bill, but if the customer asks for a breakdown, I can show it. Amounts for materials and supplies are what shows on the bill. Materials are used to fix or add to a job, such as screening, insulation and chimney caps. Supplies are defined as consumed or removed after the job is complete, for example scaffolding rental, large drop cloths, and customized fixtures made for the job.

There is no magic formula for pricing. If you can accurately estimate time on the job (including travel), then you should be able to set a profitable hourly rate that your customers can afford, and you feel comfortable with. Do not be afraid to raise or lower price based on careful evaluation of the situation. I personally feel that my knowledge and skills are worth the price. I will take less if the job is a learning experience or if there is long term or volume work. Example: I normally charge $30 to gas a groundhog den. But, tree farms get a rate of $20 per due to the volume. If a customer needs service but can't afford it, I may offer them time payments or a barter. You must be the judge.

I'll charge more if the customer did not tell me all the facts. Customers wanting service outside normal business hours are charged at a higher rate.

If you do not feel comfortable with a price, ask yourself why? If you have everything to get the job done as a professional, there is no reason to get short changed. You can turn down jobs by purposely quoting on the high side. If you sense that the job or the customer would be a pain in the butt you would rather not deal with, quote a ridiculous high price (which is how "termite pest control" companies price wildlife problems). If the customer accepts, you will at least be compensated.

Customers with large or on-going jobs are best served on a contractual basis. Try to think of all that could happen when writing a contract. If the customer has a lawyer involved, you may want one on your side to review the legal wording of the contract. You want as much in your favor as can be written in and agreed to. Once you get to the contract stage with a customer, the sale should be in the bag, so try not to be overly aggressive or do anything to cause them back out.

Following is a sample contract:

PVE Corp. **Oct. 8, 1986**

Work Agreement
between
PVE & XYZ Camps

It is agreed that:

(1) A mutually agreeable written work statement is required.

(2) PVE will provide reports showing hours worked and description of work performed.

(3) If either party is dissatisfied, the other party will be informed so any result or dollar difference can be resolved before becoming major problems.

(4) Hourly billing rates will be $27.50 per hour.

(5) Travel will be billed at the hourly rate plus $0.22 per mile from Pennsburg, Pa.

(6) PVE certifies that a minimum of $100,000 liability insurance policy is in effect, and that all workers are covered by Workmen's Compensation.

(7) Terms of payment are net 30 days upon receipt of invoice. Unpaid ________balances after 30 days will be charged 1% interest per month.

(8) This agreement carries a one year term. It may be cancelled by either party with written notification.

Agreed to by Signature ______________________________
for PVE Date ____

Printed ______________________________

Agreed to by Signature ______________________________ for XYZ
Date ____

Printed ______________________________

Some WPCO's are able to charge consulting fees. Consultants get paid to design solutions to difficult problems. You will soon find out that everyone wants to know what you know, for free. The consultant provides this valuable information for a price.

You do not become a consultant overnight, except in your own mind. Being recognized as an old hand at anything usually comes from dedication and hard work. Being creative and having ideas that work also helps. For example, someone will soon come out with a 900 number where the caller pays for answers to nuisance wildlife questions. Who will be the first?

Planning

There is good potential for lasting success in Wildlife Pest Control for hard workers who follow a plan. The typical business plan is usually an unrealistic dream sheet for attracting potential investors. What I am suggesting is something practical, a blueprint for how to get where you want to go, how much money it will cost, how much money is available, how long it will take and the resources needed. Prepare a simple organization chart, then fill in the details that go along with each block. A common sense plan will include the things you need to achieve your goals: materials, tools, time, help, etc. The plan can be formed over time, and it should be adjusted on the go, as you compare what you planned with what actually happened.

If you have ever run a trapline, think how you went about it. You "managed" the whole endeavor, and you had goals. The details were worked out before any traps were set. Scouting for animals to trap, practice making sets, equipment preparation, obtaining permission, are all out of the way before the season starts. You have the necessary gear and can make sets that work for you, in minimum time, and be on to the next location. Trapline success results from learning, studying, observing, practicing, and doing. Success in business is no different.

Ethics

WPCO's are not under much public scrutiny at this time. Most states have no government regulation. This will probably change, because the growth rate of the industry will eventually lead to more public exposure, government examination and inevitable regulation. WPCO's can control their own destiny by individually conducting their business with the highest ethical standards and forming alliances with other WPCO's who share their beliefs. Check out the associations listed in the appendix, get some information from them, then decide which groups best represent your goals. By joining and participating, you will be in exposed to new information on techniques, research and developments.

Expect the animal rights movement to attack the WPC industry, using false statements, emotional appeals, and exhibition of any bad apples to be found in the WPC barrel. As a group, we have to agree on what is ethical, and continuously refine the definition. What in the hell does this mean you ask? It means examine what you are doing, with a critical eye to improve and change, even when it is hard to do. There will be some unethical WPCO's. Do your best to distinguish yourself from them by your actions and conduct. A bad reputation is not easy to hide in this small world.

Chapter Two

Marketing, Sales and Promotion

Customers are the most fundamental need of any business. Marketing is what you do to find customers, and find out some basic information about who they are, what are their needs, and how much they might pay for your services, where they are located, and how many of them are out there. If the market is not large enough to provide steady income, the business will fail. Or, maybe there is a market, but it was not discovered or exploited. It is like scouting for fur pockets (places where fur bearing animal populations are dense), sometimes they pop up in unlikely places. Once discovered, it is time to harvest until the competition catches up. Traditional marketing taught in business schools is only useful if we have lots of time, money and help. Low budget marketing can be successful if you use less expensive tactics like effort, hustle, thought, planning and common sense. It is good strategy for a start up business to invest the bare minimum, and make a concentrated effort to discover:

1. Who and where is your market?
2. How big is it, in terms of people and dollars?
3. Who is the competition and do you have the guts to beat it?

4. Is there room for you, and is there room for growth?
5. Do you have the resources (mental, physical, financial, etc.) to perform the service?
6. What is your price, is it competitive, is it profitable?
7. How will you get customers?
8. Where you want to be in five years? (or two years or next week)

Your knowledge of local wildlife, the kinds of problems it causes, and who are the typical victims, are the first three pieces of the puzzle. Make a list of all the species you know have caused local problems. The list might look something like this:

1. Feral cat - mobile home parks, campgrounds, apartments.
2. Raccoons - business district grain storage areas, residential chimneys, attics.
3. Geese - golf course, private ponds.
4. Pigeons - municipal buildings, power plant.
5. Skunks - causing problems in new housing development.
6. Squirrel - private residence crawl space and chimney.
7. Deer - pose hazards in enclosed areas (airports, horticultural preserve, etc.)
8. Muskrat - golf course, private ponds, eating bulbs and digging holes.
9. Coyote - have killed livestock, or pets.

Now, rate each problem for dollar value per job, and how often the problem occurs. You may only get one municipal pigeon exclusion job in a year, worth say $1000, and fifty squirrel jobs at $50 (total $2500). It is better to have a bunch of little customers than a few big ones. It is essentially a numbers game.

Be honest about your ability to get the job done. If you have opportunities to remove whitetail deer with tranquilizer darts, you better have the skills needed for safe and humane handling of large animals, along with knowledge of using the darting devices, tranquilizers, and other tools. Admit what you don't know, and if relevant, go out and learn it. Books (like this one) will take you to a point, but after all the reading you have to go out and get dirty, make some mistakes, and have some real fun.

In each situation on your list there will be a contact for you to find, either an individual or a group. The contact must be the one who decides to pays for your service. In the case of the homeowner, the decision maker is obvious. However, the pigeon problem may be a matter for the town council to act on. Go to a council meeting and present yourself. Your chances of landing a job are better in person. It also helps to have references who will vouch for your skill and credibility. References can be developed by making contacts with business, church and civic groups, and volunteer organizations within your territory. Your goal is to get your message to as many people as possible, with a minimum of expense. Of course, satisfied customers are the best references. Use them as much as possible.

Problems with managed properties are a headache for the property manager. Call or visit the property, find out who is in charge, then introduce yourself and your service. Always put yourself in the shoes of the person you are trying to sell to. Remember that you are a stranger trying to sell something, and the normal reaction is suspicion. Think of how you feel when faced with an unsolicited sales call to buy light bulbs for the needy, or some other unfamiliar product or service. It can get mighty lonely when you are trying to dig up some business. Rejection goes with the territory, but it all gets better when you finally connect and are welcomed with open arms and a gasp of relief. How many mice does a fox miss before its belly is filled?

The best approach when establishing your reputation, is face to face contact with prospective customers. The best prospective customers are recommended by previously satisfied customers. It's like building a wall, one brick at a time. Absolutely believe in what you are doing, to the point of shouting out who you are and what you do from the roof tops. Your purpose is to make sure that everyone knows who to call for wildlife problems. It will take time, effort, persistence and expense to test whether you can survive.

Right now, Wildlife Pest Control is a cottage industry, with most of the people involved part time. The industry is growing at a nice pace, as evidenced by more full time operators, a franchise, trade associations, publications, articles and other signs of a healthy new business opportunity. Those who are lucky enough to recognize this are in position to take action. Taking the plunge is a risk, a rush, a race, and a great way for a trapper to live. I suggest that you concentrate on the lifestyle, not how wealthy you are going to become. Operate to please yourself and your customers, and the money will follow. One more thing: it is a vast improvement over working for someone else.

Promotions

Promotion is putting yourself first. The most valuable asset to your business is you. To deliver your essential service to the community, they must be made aware of your existence. Promotion then, is the process of getting your message out. Don't hesitate to let the local newspaper know about your exploits. Newspapers are hungry for news, and wildlife topics always make good copy. Readers are hungry for information about the wildlife they see on their properties. We all have interesting stories to tell. Be very careful in how you deliver your message to the media. Do not expect a reporter to understand your business, even if the head is nodding and the lips are smiling. You may be in for a shock if your story is misinterpreted, or you are put in a bad light. Speaking before community groups (Lions, Rotary, Kiwanis, etc.) is another way to present your service, and you may get a free meal at a dinner meeting. If you have the confidence to speak, prepare yourself well in advance. Get a good book on public speaking, take a course, or get help from someone experienced if you have never done it before. The key to a good presentation is to know your material inside and out, and practice, practice, practice. A structured presentation that you know inside and out will keep you out of a jam if distractions cause an interruption, you have a bout of stage fright (every speaker gets this to some degree once in awhile), or questions from the audience throw you off track. Time how long it takes to deliver your presentation. Each speaking situation can be padded or pared depending on how long you have to speak. I can deliver my program from between fifteen minutes, to one and one half hours.

Promotion is free advertising. For example, make announcements for your service to put on public bulletin boards. Have phone number tear-offs at the bottom of the piece. Bulletin boards are everywhere, just make sure commercial postings are permitted. Ask your friends to post them at work.

Look closely at the following sample. Did you notice that the phone numbers don't match? The lesson here is proof read all promotional material, or better yet, get someone else to proof it for you. Copy mistakes are unprofessional.

Wildlife Problems ?

Humane Removal of Animal: **Squirrel, Groundhog, Bat Skunk, Raccoon, Snake, Pigeon, Rodent, Wasp, etc.**

Fix Problem Permanently: **Screen, Seal, Exclude, Insulate, Repair, Clean: Chimney, Crawl Space, Out Buildings, Eaves, Roof.**

Comprehensive Services: **Rabies Specimen Processing, Dead Animal Removal, Wildlife Management Tools and Equipment rental and Sales, Contract Services, Consulting**

***** FREE Estimate by Phone *** Fully Insured *** Licensed Wildlife Pest Control Operator *** Full Time *** For a Solution to Your Problem:**

CALL TOLL FREE

Phil Nichols **1-800-924-7468**

WILDLIFE PEST CONTROL
Phil Nichols 1-800-724-9468

WILDLIFE PEST CONTROL
Phil Nichols 1-800-724-9468

WILDLIFE PEST CONTROL
Phil Nichols 1-800-724-9468

WILDLIFE PEST CONTROL
Phil Nichols 1-800-724-9468

WILDLIFE PEST CONTROL
Phil Nichols 1-800-724-9468

WILDLIFE PEST CONTROL
Phil Nichols 1-800-724-9468

WILDLIFE PEST CONTROL
Phil Nichols 1-800-724-9468

WILDLIFE PEST CONTROL
Phil Nichols 1-800-724-9468

WILDLIFE PEST CONTROL
Phil Nichols 1-800-724-9468

WILDLIFE PEST CONTROL
Phil Nichols 1-800-724-9468

Sample With Tear-offs

Visit local and state police stations, and talk to the dispatchers. Make sure you leave them with a way to contact you, either a business card, or letter that details your services, or both. Make contact with them at least once every six months. Be sure and thank them for any business sent your way. Other possible targets for your promotions:

- county agricultural agents
- real estate agents
- mobile home park managers
- campgrounds
- summer camps
- veterinarians
- SPCA and animal shelters
- pest control (many want only the insect business and will refer wildlife business to you if you ask them to)
- schools (all manner and kind of schools)
- golf course manager
- any large block of land that has one owner (either business, non-profit organization, individual, or some other entity)
- nurseries
- cemeteries
- dog control agents

Here is a sample promotional letter:

WILDLIFE PEST MANAGEMENT SERVICES

I provide full time service of solving problems that occur between wildlife and people. Problems happen when wildlife threaten health, safety and property. I am licensed and recommended by the Pa. Game Commission, and have been in this business since 1986. I have the necessary education, experience, tools, innoculations and desire to provide solutions that are effective, safe, affordable, humane and permanent. Please keep my card on file. I would appreciate any referrals you wish to provide.

Typical Problems:

- *skunks digging grubs in turf.*
- *skunks living under shed, porch, slab, etc.*
- *groundhog diggings destroy landscaping, vegetation, gardens.*
- *raccoons living in barn, chimney, house, etc.*
- *raccoons raiding dumpsters, livestock feed, etc.*
- *raccoon, skunk, groundhog, etc., displays unusual behaviour, such as no fear of humans, staggering around in the daytime, aggression, etc.*
- *muskrat digging into earth dams, around pools, etc.*
- *animal gets into living area, can't get out.*
- *find and remove dead animals from private property.*
- *problems with bats, snapping turtles, bees, wasps and hornets.*
- *pigeon and goose problems.*
- *feral cat problems.*
- *rabies specimen testing (processing, documentation and delivery to state laboratory).*

Typical Solutions:

- *catch, and remove offending animal in a humane cage trap.*
- *repair animal damage to soffit & fascia, prevent further damage.*
- *install chimney caps.*
- *construction of pest proof barriers.*
- *handle unusual, difficult, situations.*
- *design permanent remedies for chronic problems.*

PHIL NICHOLS 1006 LAKE LANE PENNSBURG PA 18073

215-679-8904

Sample Letter

Advertising

Money spent on advertising is a necessary part of doing business. Your goal should be to get the greatest number of good sales leads, from a minimum of advertising dollars invested. There are many ways to advertise a service, and the cost to you is like any other business cost: how good are you at bargaining? Some places where you can spend money on display advertising:

- purchased newspapers
- give away publications
- coupon books
- business directories
- telephone directories
- association directories
- non-profit group newsletters
- calendars
- programs for dance, play and recital events

Consider new ways to invest your hard earned advertising money. Directories (for example the yellow pages) are excellent because the advertisement is always there, whereas newspapers get thrown out after one reading. I place classified ads - when it's time for the rash of annual skunk problems. In this case the timing is right. Be sure to proof all advertising copy before the insertion, and even if the copy does not change between insertions, proof every insertion prior to press time.

The majority of the WPC problems in my area consist of raccoons, skunks and squirrels. The most profitable jobs involve getting these culprits out of structures, be it chimney, crawl space, attic, etc., then doing some exclusion work to prevent further problems. It's easy to convince the owner that prevention is better than playing host to defecating, destructive and possibly diseased animals. These are the people I want to reach with my advertising.

My experience is that the least expensive, and most productive advertising is by word of mouth. Your reputation is the key to a steady stream of referrals. Do not be shy about asking your satisfied customers for referrals. Customers who belong to groups, associations or networks, will know peers who have the same problems. Plug into just one such situation and you will be amazed at the volume of business that comes your way with no advertising expense at all. I have had up to ten skunk jobs in one neighborhood through referrals.

What about direct mail? Direct mail, some call it "junk mail", is costly. The expenses are roughly as follows for every piece, under one ounce in weight that you send out. Keep in mind this does not include the design, artwork, and making of a camera ready proof for the printer:

1A. Postage if first class costs $0.29 each.

1B. Postage if using bulk rate costs $0.19 each, minimum 200 pieces, all the same size, weight and sorted by zip code. The bulk mail permit cost per year is $75.00

2. Printing a one page brochure will cost between $0.15 and $0.30 depending on the printer you choose and the options such as colors, etc.
3. Mailing list cost vary between categories, starting at around $0.045 per name, with a $150 minimum.
4. If you hire out the work to actually do the mailing, it will cost around $0.08 per piece.

The cost then to reach one potential customer with your message is between $0.38 and $0.75.

Here are some sample categories for mail lists, with the quantity of names that are available (on labels) in parenthesis:

- City Managers (3,900)

- Children's Camps (5,000)
- Cemeteries (5,500)
- Camps, Sports & Recreation (5,000)
- Campgrounds (6,100)
- Pest Control Services (17,000)
- Government Public Health Offices (7,000)
- Municipal Park & Recreation (2,300)
- Municipal City Manager (4,100)
- Municipal Police Chiefs (6,500)
- Farmers Co-op (7,400)
- Farmers, wealthy (73,000)
- Executives, Agriculture, Forestry, Fish (40,000)
- County Agricultural Agents (10,200)

Names from any category can be purchased by zip code or other sorting parameter. Get list catalogs from several sources and use your imagination to get ideas. Even if you don't do a mailing, these list catalogs will stimulate ideas for places to promote your service in less expensive ways. Bulk mail is expensive and must be done right to get a decent return.

Mailing List Sources:

Karl Business Lists
1-800-367-4298

Research Projects Corp.
1-800-243-4360

Hugo Dunhill Mailing List Catalog
1-800-223-6454

Alvin B. Zeller
Mailing List Catalog
1-800-223-0814

What you send out might get looked at if it grabs their attention. Most likely it will end up in the trash. You can expect ten to twenty returns per thousand sent out. What you offer must be a product or service that appeals to the needs and desires of the group you chose to target, and the money you take in from these returns must at least cover the expense of the mailing. Bulk mail can be a gamble, but it does work as you can assume by the contents of the average mailbox.

A sample bulk mail piece (tri-fold) is illustrated on the following two pages. It was an unsuccessful mailing. I purchased a mail list of people over 55 years old who owned homes in nearby zip codes. I purchased a list of 4,500 names, and mailed out 1,000 of them as a test. I got two sales from the mailing and lost around $450. But now I have a bulk mail permit and am getting all excited about the harebrained ideas I've been having for the next mailing. This is the cost of education. I include my experience here as something to ponder by anyone who considers doing a bulk mailing.

ARE YOU RUNNING A CRITTER MOTEL?

WHAT IS LIVING UNDER YOUR SHED?

P V E
PHIL NICHOLS
1006 Lake Lane
Pennsburg, Pa. 18073

1-800-724-9468
in Pa. 1-215-679-8904

BULK RATE
U.S. POSTAGE
PAID
RED HILL, PA
PERMIT NO. 47

FROM ____________________

____

____

PLACE
STAMP
HERE

PVE
P.O. Box 84
Pennsburg, PA 18073

Sample Brochure : Tri Fold Front

IF YOU WANT TO CLOSE YOUR CRITTER MOTEL FOR GOOD . . .
CALL 679-8904

FOR MORE DETAILS AND A **FREE ESTIMATE,** OR DETACH & RETURN THE CARD BELOW FOR MORE INFORMATION.

FEATURES

- *INEXPENSIVE*
- *SANITARY*
- *PERMANENT*
- *HUMANE*
- *GUARANTEE AND WARRANTY*
- *GAME COMMISSION WPC #0232 RECOMMENDED*

BENEFITS

- *GUARANTEED TO EXCLUDE ALL BURROWING AND TUNNELING ANIMALS, ALL REPTILES AND ALL RODENTS.*
- *NONE OF THE UNFORTUNATE VICTIMS OF THE RABIES EPIDEMIC (Racoons, Skunks, etc.) CAN LIVE AND BREED UNDER YOUR SHED.*

WHY TAKE CHANCES?

PVE specializes in inexpensive, sanitary, permanent and humane pest control methods. We can eliminate your pests, just give us the opportunity. **Send this card in TODAY!**

❑ *I would like a FREE proposal, please call me for an appointment.*
❑ *I would like to have more information.*
❑ *Please keep me on your mailing list.*

Name ______________________________

Address ______________________________

City ____________________ State __________ Zip __________

Phone ______________________________

Sample Brochure Tri Fold Back

There is an old cliche "it takes money to make money". How much does it take? The traditional business comes up with an advertising budget. The amount of money you have to spend on advertising, or anything else for that matter, has to be spent in the right place, in order to get customers. Always remember that customers are the lifeblood of any business, and the secret to success is to get lots of them, find out what they want, and give it to them as quickly and cheaply as possible, while ending up with lots of profit. Advertising methods must be changed if they do not work. When a method works, keep using it.

Chapter Three

Skunk Problems

The common striped skunk is in every state except Alaska. We all know and love their stinky juice. From a cloud like spray on close targets, to a stream, or drops, shot at victims up to ten feet away, it can shoot at multiple targets at different angles and distances, whirling in an arc, pulling the trigger as required. The thick, amber colored chemical (mercaptan) has a sulfur base, and its' discharge is controlled by squeezing muscles that surround the anal glands. The value of pure skunk essence ranges between five and twelve dollars per ounce, wholesale. What is it used for? Deer hunters use it to hide their human odor. Trappers add it to lures because it keeps working in sub zero temperatures. It is also blown through pipelines to discover leaks. And, the perfume industry uses it as a base element, after some molecules are rearranged!

The skunk pelt has a beautiful, glossy sheen, and is still bought and sold in the fur industry. During the great depression, it was possible to earn significant money catching skunks. This changed when the "truth in labeling" act caused the devaluation of skunk pelts. The act dictated that skunk garments to be labeled "skunk". No longer could the beautiful and functional fur coat be called "North

American Sable". It had to be labeled "SKUNK". The latter does not make the same fashion statement as the former. Today, skunk pelts are used in a variety of niche markets. The value of raw skunk pelts ranges from one dollar to eight dollars. I read somewhere that skunk meat tastes like veal. If you can verify this, please give me a call, I would like to talk with you.

Skunks are in the same family as weasels and minks, and have the same type of thin body structure. This allows them to squeeze through small openings. I heard a story about a skunk that tried to get into a dumpster through the drain hole, but then got stuck. A kindhearted but witless person tried to free the animal by smearing Crisco around it's head. The good Samaritan received his thanks from the other end of the appreciative skunk.

The majority of provoked skunks will give some warning before spraying their victim. However, a small percentage are renegades that don't need a reason to shoot. I hope you never have to deal with such a creature. "Good" skunks give three signals: the first is stomping of the front feet. This may be combined with clicking of the teeth, hissing, or a noise that sounds like a combination of gargling and growling. On some surfaces, the foot stomping routine produces a loud noise. The second sign is lifting the tail and pointing its' butt towards the target. Finally, the skunk will turn it's head to look at the target, while arching its' back. At this point the target is in sight and the trigger is being pulled. What is the safest course of action if you and a skunk have a close encounter? Freeze where you are and observe the signs the animal gives. If it is agitated and lifts its tail, remain calm, talk to it using your "poor, poor, pitiful skunk" voice and the animal will calm down. You can slowly back off if it hasn't started foot stomping or tail lifting.

Skunks settle in to a home range roughly equal to two square city blocks. This is where they live, but they may have many places to hole up at the end of their nightly activities (they are nocturnal). Males may roam outside of this range, and may get caught wandering around in strange territory after the sun comes up. Forced to seek temporary refuge wherever they can find it, they will duck into any likely looking place. "Is there a garage door open?" In late summer and early fall, juveniles searching for a home (either because of wanderlust, or mom kicked them out), will have similar inclinations. Juveniles can squeeze through a very small opening, just a little larger than the diameter of a tennis ball. Skunks look fluffy because of their fur.

Any skunk found wandering around in the daytime should be treated with caution since they are susceptible to rabies and other diseases. A majority of rabid animals simply die unnoticed, with very few getting to the aggressive, foaming at the mouth stage. A call to deal with a slobbering psychopathic skunk will really put your resolve to the test. Still want to be a WPCO?

Skunks will stay with a favorite food until it is all gone, then will look for a new source. In the spring and summer they feast on an abundance of insects. Skunks are not bothered when digging into nests of stinging, biting bees and wasps to feast on the larvae. During this season, soft, juicy grubs are the favorite forage. Skunks dig divots in manicured lawns, searching for Japanese beetle grubs that are just below the surface. They also consume vegetable matter such as fruits, various grasses, berries, nuts, mushrooms, tree buds and garden vegetables. They will scavenge during hard times, eating what is available to survive.

Skunks eat meat, and will devour fresh, not so fresh, rotten and really rotten, newts, salamanders, snakes, small rodents, birds and road kills of just about anything. Skunks will raid bee hives and chicken coops (for eggs and chickens). Unsecured garbage left outside will be investigated by skunks. Most of the skunks eating activities are beneficial to man, because they eat the rodents and insects that cause us problems.

Male skunks start looking for females to breed with in February, and mating continues until the end of March. Litters are born in May and June. The number of young varies from two up to twelve. They stay with their mother into late summer and early fall, following her everywhere. Mother skunk has some way of recognizing when her babies are mature enough to go it alone. She kicks them out and they wander, looking for a place to call home. This dispersal is an annual event that creates opportunities for WPCO's. All creatures must be weaned. Do some research, and find out the times and patterns for dispersal of all the species common to your area. Teenage skunks may find refuge under porches, sheds and decks, and foundations. They can easily dig out a small opening to get under a structure. They will provide you with an entire neighborhood of problems.

Skunks den in old groundhog holes, hollow logs, and under buildings, abandoned automobiles, porches, concrete slabs, and outbuildings. During the coldest part of winter, they may stay in the den until a warm spell, but are not true hibernators. Den sites vary with the season. In the spring and summer, radio collared skunks tended to bed above ground in buildings and croplands. In fall and winter they moved to underground retreats, and will stay there during bouts of bitter cold.

A lawn that is pock marked with two to four inch diameter holes, is a sure sign that skunks are digging for grubs. Another sign of skunk digs is a larger hole where they have uncovered an underground wasp nest. The dig will be as big as a football, with pieces of the honeycomb scattered around (minus the larvae), and some wasps will be buzzing around the destroyed remains of their nest. When raiding sweet corn, the bottom of the ear is eaten. When raiding chicken coops, they open the egg at one end to suck out the contents, and may kill one or two chickens.

PROBLEM: Skunk falls into window well, can't get out.

The animals habit of following a line (fence line, wall, hedge row, etc.) sometimes results in their falling into a ground level window well. This may not be discovered until after the animal has died, and the body is starting to decompose. Leaves and debris may hide the carcass. When the scent glands decompose, the odor will be released, and it will be very strong and lasting. Where there is a persistent odor of skunk and the source can not be found, check all window wells thoroughly. If the animal is alive, put a skunk ladder down into the window well. Make sure the angle is not too steep, so the skunk can climb out. The skunk may be alive, but for some reason the ladder will not work (too weakened to climb out, or animal must be removed immediately). In this case use a syringe pole to inject six cc's of euthanizing fluid (see appendix for source) into the chest area.

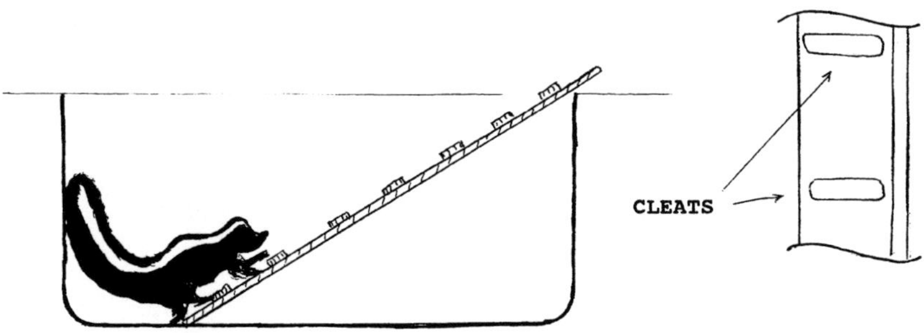

SKUNK LADDER IN WINDOW WELL

The Skunk Ladder.

Approach the window well with as little noise as possible, during mid day, when the sun is out. The animal will most likely be curled up asleep. Place the board in the well then leave. The skunk will wait until dark to climb out.

If the animal is a baby (about the size of a kitten), it can be lifted out by hand. I put a leather glove on and grab them by the skin on the back of their neck, then put it in a burlap sack. Because their scent glands are not developed, the spray is very weak, the volume is tiny, and the odor doesn't have the same force as mature skunk essence.

If the window well needs deodorizing, remove all debris, leaves, sticks, loose dirt, etc. (wear disposable rubber gloves). When the well is thoroughly clean, pour deodorant all over the area. Try 3% hydrogen peroxide or a commercial deodorant. Leave it for a day, then reapply if needed in twenty four hours. The odor may emerge again after rain. To prevent skunks from falling in again, install a barricade, or screen. Or, the skunk ladder can be left in the well.

PROBLEM: Skunks digging up lawn.

Skunks will dig for grubs and larvae (for example, the Japanese beetle grub is a favorite), creating divots the size of golf balls in the turf. These pock marks in the yard can be spread out or concentrated in one area. The solution is to remove the culprits, using a suitable box trap. Box traps for catching skunks, and where they can be obtained, are listed in the appendix. If the trap has a wire bottom, set it on a piece of plywood. This will stop the skunk from digging up ten pounds of dirt which it then brings into the cage. This makes a hole in the lawn, and makes the trap real heavy when you have to carry it.

Before you buy traps, think about how they will be used. There are box traps designed to catch skunks, made from solid material so the skunk can not see out. The havahart is a multi purpose trap with solid doors and open sides. Remember, skunks don't spray what they can't see.

Put the bait in the rear of the cage when using a single door trap, or in the middle of a double door trap. Use an apple, marshmallow or some other sweet morsel, something that free roaming cats will ignore, and something that will not turn rotten and be covered with ants in warm weather. If cats are not a problem, a fishy flavored dry cat food works good. Along with odor, the bait used should have some eye appeal, which can be enhanced with a piece of aluminum foil.

Make sure the trap does not wobble by placing shims (sticks, pebbles, etc.) underneath. The trap should not wiggle or move when the animal enters it. Dry fire the trap a few times to check the mechanical workings. Look for places where the moving parts might stick. Friction could cause the door to close slow, or not at all. The Victor Havahart traps have a locking bar that must be set in position.

The best location to set the trap is close to where the skunk is digging. Results can be expected in, from one to three nights. However, animals get into a variety of predicaments. Getting from one side of the road to the other is a one hazard. Great horned owls will prey upon skunks, they don't seem to mind the odor. In cold weather the target animal may snooze until the weather moderates. Or, the animal may just decide to go somewhere else temporarily.

What do you do when the skunk is caught? Remember, the skunk will not spray what it can't see. A trap made of a solid material can be picked up and taken away. I once dropped a trap with a skunk inside, and it did not spray.

Choose a spot to release the animal where it will not become a problem for someone else. It is best to have a watercourse (river or stream, reservoir, etc.) between the old home and the release point. If the skunk is caught in a Victor Havahart trap, quietly approach the end with the solid door, so the skunk cannot see you. Pick up the trap using the handle and keep the trap straight out in front.

Pickup Skunk

Skunk In Truck

Park the truck as close by as possible, with the gate down. Put down a washable or disposable tarp, to keep the vehicle clean in case the animal poops or pees during the ride. Slide the trap in and gently close the gate. Then, take the skunk for a ride. Take it easy and you will have no problems. A pickup is preferable, but any vehicle will do if you remember to take it easy, and do not let the skunk see you. Release the animal as described before. To open the Havahart trap, unlatch the locking bar from the door you want the animal to exit from, and jam the other door's locking bar with a pencil or stick.

Point the end of the trap away from you, toward some sort of cover that the skunk can see when it exits, such as high grass, a field or woods. Then tip the trap upside down and move away. The skunk may take a minute or so to leave.

An additional option is to kill the animal. In most states skunks are classified as furbearers, and are protected by law. Your local game warden or conservation officer can advise you. There are rabies epidemics in many areas, and skunks are primary carriers. The state may require WPC operators to kill all captured skunks since there is no way diagnose the disease in a live animal.

There are many ways to kill animals in cage traps. It can be drowned by lowering the entire trap into deep water. The animal will probably spray. Methods that do not result in spraying vary in difficulty and effectiveness. The syringe pole is a method I have not used in a cage trap, but I have been told it is a method that definitely does not cause spraying. I have injected skunks in different circumstances and there was no release of odor. My philosophy is to get the animal away from the site as fast as possible, since I guarantee "no odor".

Some other ways to kill skunks: the animal can be put in a gas chamber and killed with carbon monoxide. Or, you can release the skunk in an open field, then shoot it with a shotgun. I use high brass #6 shot.

If you have a situation where mother skunk is caught in the cage trap, and her family of youngsters are outside, pacing about, worried and confused, what do you do? Cover up the trap that mom is in, then catch the babies with a net or blanket, then put them put them in a carrying cage or burlap sack.

Another way is to place more cage traps around moms covered trap until the entire family is captured. The opposite situation has mom outside and the babies inside the cage. To catch mom, leave the caught babies where they are and set another cage right next to them.

PROBLEM: Skunk living under a structure.

In the wild, skunks den in hollow logs, caves, rock piles and abandoned groundhog holes. "Town" skunks den under our sheds, porches, ground level crawl spaces, the hollow places under cement slabs, wood piles, abandoned automobiles, etc. If a location shows promise, the skunk digs to expand the hole and create a living space where it will be dry and secure. How does it know there is a hollow spot under a cement slab? I found where an underground wasp nest next to a cement slab, was discovered and eaten by a skunk, which then dug out the initial cavity to create a den.

Skunks may have a number of dens in their range. They will use the one closest to where they end up after a nights scavenging. If a food source is plentiful, good tasting, and close to home, the skunk will use the den until:

1. den is destroyed or made inhospitable
2. food source is exhausted
3. the skunk dies
4. the skunk moves on for no apparent reason

Skunks do not usually spray in their den because the stench is just as bad to them as it is to us. Odor at a den usually is the result of a surprise attack. Dogs probably account for most of these attacks in developed areas. My Labrador Retriever sticks his face in every skunk butt he can find, never seeming to remember the results and after effects of his previous encounters.

There are a couple different approaches to removing skunks from around structure. If it is one animal, it can be caught in a cage trap as described previously (skunks digging up lawn). It can also be caught in a conibear trap placed over the opening at the structure, which probably will result in odor. Be sure there is enough clearance for the jaws to close. Use the 120 size with two springs, or the 160 size. Stabilize the trap (see photo) with cross sticks and fasten it solid. Conibears kill any animal that passes through the jaws, and can damage human fingers and hands. Where these are even remote possibilities, either don't use it, or construct a cover to prevent children or pets from harm. I use a weldwire cover, shaped like a covered wagon top, two feet long and staked solid. You may want to display a warning sign for the curious:

***** WARNING: TRAP SET *****

Conibear

Determine how many skunks are present. The animal(s) leave the den any time between dusk and midnight, usually around the same time every night (weather permitting). The homeowner may be willing to join in the fun by watching (from inside the house if possible) and noting the time when, and how many animals leave the den. Skunks are normally loners. The exceptions are a mother with babies; and, in extreme cold, more than one animal may share a den for warmth. Spread a layer of sand, flour or similar substance at the den entrance, to reveal the tracks of the skunk(s) as it leaves. Check for tracks periodically starting at dusk. After the den is empty, block the entrance.

Keep in mind that from May until August, there may be babies left behind when mother skunk leaves. If the number of animals contained in the den is uncertain, rig a door made from weld wire, hinged to open one way, so the animal can push it open to leave, but not get back in. As before, the entrance should be dusted for tracks. If no tracks are evident after a few days, seal up the entrance.

Excluding Skunks: General Procedure For Installation Of Barriers

Creatures that enter buildings can cause serious damage to insulation, wiring, and other structure. All are expensive to repair. The best fix for these kinds of problems, after the creatures have been removed, is to install a permanent barrier. The material is a corrosion proof mesh screen, buried in a trench below ground level, with the above ground end fastened to the structure. The trench is filled with stone, which discourages digging and entry by groundhogs, skunks, field mice, rats, and voles. This barrier will exclude all digging pests as long as the directions are followed in detail.

TOOLS: shovel, hammer, tin snips, pliers, hacksaw.

MATERIAL: #3/4 galvanized staples, 26" x 96" x 1/2" mesh galvanized lath, 3/8 x 8 foot REROD cut to lengths as needed, 3/4 inch stone as needed, 14 gauge galvanized wire as needed.

DIRECTIONS: Dig a trench around the perimeter of the structure, one foot wide and one foot deep. see Exclusion barrier, side view).

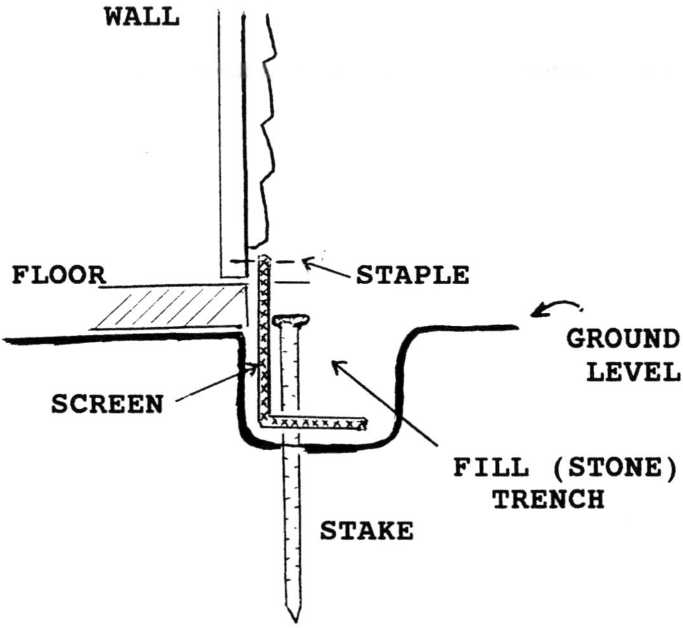

Exclusion Barrier -side view

Cut the lath to fit from the base of the structure, to the bottom of the trench, allowing for an outward bend. Put the lath in the trench, and fasten its' top to the base of the structure, using the 3/4 galvanized staples . If the structures base is not wood, use a fastener appropriate for the material, such as sheet metal screws with washers, into aluminum siding. Next, bend the lath at the bottom of the trench, out and away from the structure. Pound stakes every two feet along the perimeter, positioned to butt up against the structure. The top of the stake should be at ground level when done pounding in.

Chapter Four

Raccoon

Raccoon Problems

Raccoons are abundant throughout the lower forty eight states. When they are threatened, cornered or challenged, they will use their teeth, claws, and strength to counterattack, no matter how big the opponent. Like humans and monkeys, they can be playful and curious, using the thumbs on their fore paws to grasp, hold and manipulate food, and objects to be examined. They are usually not far from water, where they hunt and explore. reaching into holes and lifting rocks, looking for fish, frogs, tadpoles, clams, and aquatic insect larvae.

Raccoons have a range of guttural sounds, squeals, chirps and growls. They growl when agitated, sometimes seem to be talking to themselves, and purr loudly when content.

Males will claim territory up to twenty five square miles, while females range somewhat less. They can be found in different sections of their territory, usually where their favorite food of the moment is available. It may be sweet corn in the fields this week, acorns up on the ridges next week, with every Tuesday evening reserved for raiding garbage cans. Mother raccoons must focus on a food source with enough volume to satisfy her broods insatiable appetites.

Normally nocturnal, a raccoon seen wandering around after sunrise has either been driven out of it's den, or is sick. Many of the diseases and parasites common to raccoons can also infect humans, pets and livestock. They get rabies, parvo, distemper and roundworms. I have never seen a raccoon with mange.

Is it true that raccoons wash their food? My opinion is that dipping a piece of rotten meat in sewer water is not an act of hygiene. Maybe it is some kind of nervous habit.

Raccoons are technically classified as omnivores (feeds on whatever is available). They snooze in their dens during periods of extreme cold, and when the temperature moderates they go out looking for food. In winter, opportunity has preference over taste. They will eat road kills, tear open garbage bags, and prowl the waterways that aren't frozen. In the spring, summer and fall, they fatten up on insects and larva, fish, snakes, baby muskrats, birds, baby birds and eggs (waterfowl and other ground nesting birds are sometimes extremely vulnerable), newts, salamanders and frogs. Urban raccoons grow fat on a bounty of nourishing cuisine such as donuts, burgers, fried chicken and pet food. Raccoons may get a taste for poultry.

My favorite baits for attracting raccoons are fresh fish, shellfish oil, shells from clams, mussels, shrimp and lobsters, chicken, commercial banana bait, large white marshmallows, peanuts, and any of the aforementioned wrapped in aluminum foil. Bait that has a lot of visual attraction will be investigated by curious raccoons. Raccoon gland lures are most valuable in mating season.

February to March is mating season for the northern raccoon, where males roam all over, checking out dens for a receptive female. Raccoons are polygamous, but the male sometimes stay with the family group until the next mating season. The average litter size is four, born in April or May.

Away from the city and suburbs, the traditional 'coon den is a hollow tree. Out in the country they also use caves and rock hollows, mine shafts, barns, silos and large groundhog holes. These places are practically non-existent in urban and suburban areas, which still have large populations of raccoons. The animal has adapted to using chimneys for hollow trees. It also uses crawl spaces, attics, abandoned automobiles and buildings, junkyards, dry culverts and sewers. During warm sunny fall and winter days, raccoons may sunbathe in trees, stretched out on top of a squirrel nest or some other comfortable high perch.

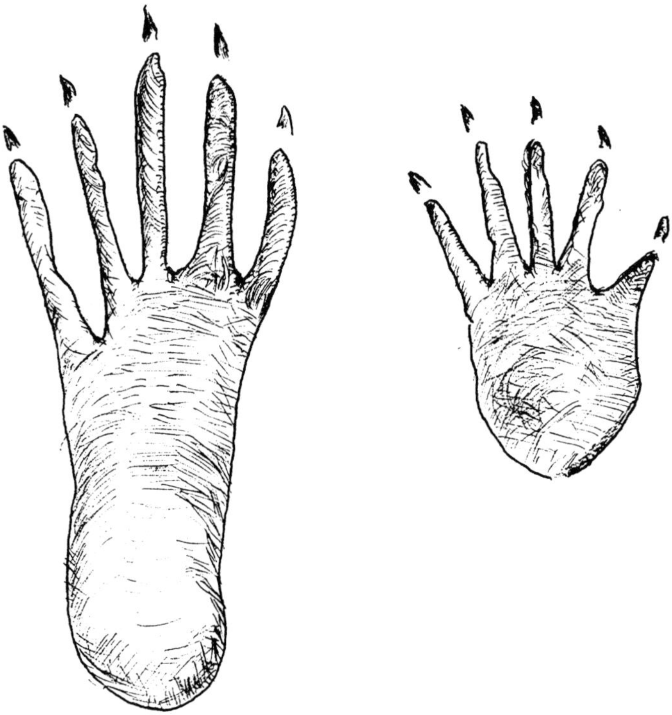

Coon Foot Prints

The raccoons' front paw prints resemble little human hands, and the rear paw prints look like baby feet. These are seen clearly in mud, soft dirt, and damp sand. Muddy prints are evidence of their climbing up smooth, lighter colored walls. Their sharp claws make scratch marks, and when the same path is used repeatedly, a distinctive trail will be worn into the surface. The adult raccoon is inquisitive and has great strength. Methods that keep most scavengers from getting into trash cans can be tested to the limit by determined raccoons. They use their strength and gripping ability to break into chicken coops and feed bins. A patch of corn can be destroyed overnight by a hungry family of raccoons. They will take a few bites from an ear then become impatient, discard the ear in hand and go rip down the next stalk. Raccoons killing poultry leave the heads and feet as evidence.

Raccoon droppings are an obvious sign of their presence. They will use the same spot as a toilet, piling up big mounds of poop, which is usually tubular, a little smaller than the diameter of a hot dog, and containing evidence of the diet. For instance, scat from eating berries and fruit will have a brownish-red color, and will be filled with small seeds. If acorns are abundant up on the ridges, the poop color will be tan to light brown, with a smooth, mealy consistency. If the diet is corn, the kernels show up as expected. Chicken feathers will be present in the turds of a chicken killer. If the contents are scales, or shells, the animal was eating fish, crayfish, or snake.

PROBLEM : Raccoon in chimney.

Homeowners can get really disturbed to find a raccoon living in the chimney. The animal may be a mother with babies, which could be anytime during the months of April through August. She can be aggressive, and there is not much space to work in. Keep the damper closed! Here is the list of things I use, yours may be different:

Materials and Supplies

- full strength ammonium hydroxide
- bactericide (Microban)
- chimney cap

- flea spray (carbaryl or methoxychlor)
- 14 gauge wire
- a roll of plastic and drop cloth
- masking tape

Equipment and Tools

- ladders (extension, step, chicken, etc.)
- flue rods, brushes, etc.
- large capture net
- choke pole
- transport cage
- 220 conibear
- hand tools
- safety equipment: glasses with side shields, face mask, gloves, safety rope and harness, etc.
- 2 HP WetVac

A sturdy ladder and stabilizer may be needed for getting to the roof, where the degree of steepness will determine how the chimney will be approached and worked on. Be aware of anything on the roof that would cause you to slip (ice, little sticks, wet leaves, etc.) Take your time. The height of the chimney above the roof has to be considered. Is it safe? Before you go leaning on the chimney, be sure that it isn't going to topple over. Put up scaffold or chicken ladders if the roof is too steep and make sure everything is secure. The job will not go well if you don't feel safe. The first part of the job is getting the animals out of the chimney, as soon after sunrise as possible.

Decide beforehand whether you need to capture the raccoon once it is driven from the chimney, or just let it go on its' way. Capturing is obviously more difficult. I rig a 220 size conibear over the flue using 14 gauge wire, bungee cords and large adjustable clamps. Or rig a net the same way. Whatever you rig up must be secure and strong enough to keep a large powerful, struggling animal right there. A raccoon can go a long way wearing a conibear. If the decision is to drive it from the chimney and let it go on it's way (providing this does not pose any danger to people, pets, livestock or property), be ready to climb up to shoo it from the top of the chimney. I use a large capture net, which protects me in case the animal attacks. If you strap a large net over the flue the coon may stay put on top of the flue, in which case it must be pushed off so it gets tangled in the net. When the animal is immobilized, it can be safely injected with euthanizing fluid. But, the first step is to drive that raccoon out of the chimney.

Remove everything from the fire box, then crack open the damper to create an up-draft, (if there is no fire box, as with a furnace chimney, a heater technician is needed at this stage). Place a pan of full strength ammonium hydroxide (available from me as Chimney Flush see Appendix) in the fire box, being careful not to breathe the fumes or spill any of this caustic solution on yourself. Have someone posted outside to watch for animals exiting the chimney and to note how they get down from the roof. Later, it may be possible to construct a permanent barrier on this path to prevent any animals from using it again. When you are certain the raccoon is gone, carefully remove and secure the ammonia, fully open the damper (or take it out if possible) and remove all debris from the flue. Use a pipe brush to get at debris that is out of reach. Wear gloves, a mask and safety glasses. Vacuum the area thoroughly if you have a shop vac. Apply bactericide, because the feces may carry disease which is transmissible to man and pets. Fleas are common parasites on the raccoon, and the flea eggs will roll off the animal, into the nest and surrounding areas. Spray the area with flea spray. When everything is cleaned up, cap the chimney. The animal may return so be sure the chimney cap is installed tightly.

Watch out for the hazards that go with this job, such as: falling off the roof; being gnawed on by an indignant raccoon; soot and assorted junk raining into your face; cuts from sharp edges on the metal chimney cap; chemicals and insecticide spills; anything unexpected.

List everything you do and why it is done on the bill:

1.	Remove raccoon from chimney	$50.00
2.	Remove raccoon nest and feces, clean chimney	$50.00
3.	Apply bactericide and methoxychlor for roundworm larva and fleas	$30.00
4.	Install chimney cap	$35.00

PROBLEM: Raccoon living in house (crawl space, attic, etc.).

This problem has all the difficulties of the previous problem. Is there a mother with babies? How does the area get cleaned up and sanitized after the animals are removed? How did they get in? Their entry way must be found and made secure. Think about degree of difficulty, your abilities and resources, then decide how to approach the job.

Find where the animal is getting in by looking for signs such as prints, a worn path (which may be very faint depending on the surface), openings larger than the diameter of a grapefruit, bits of food, garbage, or droppings. Once you find it seal it tight, after the animals have left, or have been removed with cage traps. Next, remove all nest material, offal and feces, vacuum thoroughly and sanitize the area with bactericide. Finally, spray carbaryl or methoxychlor to kill all flea eggs and larvae. Raccoons (along with many other creatures) like to use fiberglass insulation as nest material. Dress in coveralls, gloves, safety glasses with side shields, and a mask, to protect yourself from the irritating fiberglass. If you replace ruined insulation, use construction techniques that prevent future problems, by either installing a screen of one quarter inch galvanized mesh weld wire between the studs, or by using foam or plastic insulation materials, something the critters will not use for nest building.

General Procedure For Trapping Raccoons In Cage Traps

Use a cage trap designed for capturing raccoons. Other styles of trap may work, but I strongly recommend giving yourself every edge possible. Raccoons will test everything to the limit. They have taught me to do things right the first time, to pay attention to all the details, and to check everything twice. I respect the raccoon for it's tenacity, cleverness and strength. However, it will always be a simple creature more concerned with feeding its' face than safety. I am much smarter than a raccoon, uh-huh? But if I foul up some detail, the raccoon will probe and poke, and may find a way to escape. The escapee has now learned about cage traps, and may be apprehensive and more reluctant to enter the trap the next time around. Do it right the first time.

Put the trap where raccoon sign is most obvious. Keep it away from anything the raccoon could grab by reaching its paw through the wire, and get enough leverage to tip the cage over and escape. Bait the trap with something sweet and conspicuous. Those big, white, fluffy marshmallows are highly visible, and the sweet toothed raccoon relishes them. Or, use any local food item that the target animal has been eating. For example, if livestock feed is being pillaged from a container, seal the feed container so the coon can't get into it, and use the feed for bait. Put most of the bait at the very rear of the cage, and position a small sample just outside the trap.

The raccoon will investigate the outside of the trap first, and may try to pick the cage up and dump the contents out, as it would do to a trash can. Prevent this by driving a stake down through the rear of the cage into the ground, or put a cinder block or equivalent weight on top of the trap.

Check the trap daily. On most jobs the customer calls me when there is a catch. It is important to take care of the catch quickly, because no matter how strong a trap is, a raccoon, given enough time, may fight with escalating violence, until it either escapes or is exhausted. This is inhumane and unnecessary. The raccoon will stick its paws through the cage wire, attempting to pull everything within reach into the cage. Take advantage of this trait by placing the trap on top of loose hay or straw. The raccoon will occupy itself by pulling the hay into the trap. It will curl up and sleep on the nest it makes.

The raccoon is classified as a furbearer or game animal, protected by a states game laws. When a protected animal becomes a threat to people or property, the state game commission or wildlife department usually gives permission to deal with the problem, and may give instructions on what to do with the animal after it is caught. In Pennsylvania, all nuisance raccoons must be destroyed due to the rabies epidemic. The way to kill an animal in a cage trap depends on the situation. I have used drowning (as described in the skunk chapter) which is the easiest way. Use caution when picking up the cage, because this animal is aggressive, quick, and has sharp teeth and claws. Wear leather gloves, use the handle, and keep an eye on the animal. Another method is to shoot the raccoon, using a 22 pistol and CB caps. Distract the animal by waving a free hand in front of the cage. When the raccoon looks, aim right through the ear in a straight line. Be careful when shooting at close range. I have had a slug ricochet off the raccoons hard skull. Another alternative is to put the cage and animal into an airtight chamber, then introduce carbon monoxide gas.

PROBLEM : Raccoon getting into trash.

It is irritating to have to pick up your trash after the raccoons have put it on exhibit all over the street. Other critters are guilty of being trash pickers, such as skunk, possum, dog and cat. But, it is raccoons who will take over a trash site, to the point of being aggressive toward the homeowner. It is best to remove this kind of animal by trapping.

To prevent further problems:

1. Put trash bags in a big wheel fifty gallon container that has a lid. Rig a bungee cord over the lid to further secure it. You can buy this kind of container for around $75.
2. Build a square pen with a hinged lid and latch. Use one half inch weld wire..

PROBLEM: Raccoons raiding crops

Human beings are not the only creatures that love the taste of juicy sweet corn in the milky stage. Deer, skunks, blackbirds, squirrel, and raccoon all share the passion for this favorite summer vegetable. Each of these creatures have their own way of eating corn. The signs that raccoons leave when eating corn are husks peeled back, and partially eaten ears. The stalks may be pulled or broken over. The casual grower may be annoyed enough to call you for help because the garden is their pride and joy. They want to pick your brain for a cheap solution. They will be a hard sell beyond minimum service, or trap rental, because it is usually frustration rather than economic loss that prompts them to call. The best deterrent for keeping raccoons from a garden patch is an electric fence, constructed with a single hot wire on top of chicken wire. Turn it off during the day.

PROBLEM : Raccoons killing chickens.

The guilty animal can be caught in a cage trap placed inside the chicken enclosure. There is no way to keep a determined raccoon from getting in poultry pens, without enclosing the entire pen in sturdy wire. Raccoons with a taste for chicken will probe for weak spots to get in, they will climb, and they will jump from a great height. When the animal is caught, it should be humanely killed, since it will get into the same trouble if released elsewhere.

The conibear trap can be used to cover an opening being used by chicken killing raccoons. Use the Victor model 160 or bigger in this situation. Be careful setting the trap and keep in mind that the jaws will close on anything that goes through the opening. Pets and children have to be considered.

PROBLEM: Possible sick raccoon.

A raccoon walking around in the daylight may be sick. Animals with weepy eyes, that stagger, seem lethargic, or are immobile should be treated with caution. In developed areas be sure all children and pets are out of harms way before you start.

Usually, the sick animal will be easy to put a net over. Only one time did I have to chase what I thought was a sick animal. As it turned out, the raccoon was frothy at the mouth and half blind from being sprayed by a skunk.

When the crisis is over, use a bactericide on the area where any blood, or other body fluids may have been spilled. The remains should be buried or burned. If any humans were exposed, have the head tested for rabies by your State Department of Health. I charge $100 for processing the animal, filling out the paperwork, packaging the head properly, and express delivery to the state lab.

Procedure For Submitting Specimens For Rabies Examination

1. Wear surgical gloves to handle the specimen. Cut off the head at the neck. Clean up all fluids.

2. Only the head should be submitted with the exception of bats and rodents, in which case the entire carcass is preferred..
3. Do not submit live animals.
4. The specimen should be maintained at refrigerator temperature - never frozen..
5. For shipment, the specimen should be placed in a plastic bag and this in turn placed in a leak proof unbreakable container along with a sufficient amount of packaged ice **DO NOT USE DRY ICE** Avoid putting the specimen in direct contact with the refrigerant.

Chapter Five

The Groundhog

This burrowing rodent has a considerable following. In Pennsylvania Dutch country, there are so called "Grundsow" organizations. I have no idea what functions they perform, or why they are so fond of this large stinky rodent. Every February 2nd you can count on front page publicity in all the local papers, reporting whether Punxatawney Phil has seen his shadow or not. Farmers are not so fond of the groundhog, because the holes it digs cause problems, which means time and expense. Just what those lazy farmers need. Groundhog holes upset horseback riders, cause sprained legs and damage farm equipment. Christmas tree farms may be plagued by groundhogs digging at the roots.

Many woodland creatures make use of abandoned groundhog dens. Red fox spend their lives in the open, except when the female bears young. Then a suitable groundhog burrow will be expanded to make a secure, underground birthing place. Skunks, opossums, raccoons, weasels, mink, snakes, rats, mice, rabbits and even pheasants use groundhog holes to den or seek refuge in.

The groundhog is a true hibernator, disappearing from the scene over the cold winter months. It seals off the burrow with dirt, then curls up in a ball and gets energy from it's thick layer of blubber during the long sleep. When spring arrives, the males wake up and start looking for a prospective mating partner.

Woodchucks eat just about any kind of vegetation growing in their territory. After coming out of hibernation they will nibble on fruit tree bark, and the twigs and buds on low growing ornamentals. They soon switch to emerging grass shoots and other vegetable sprouts. During the summer they feed on this and that, stopping to rest and look around and sun bathe. At this pace woodchucks consume about twenty percent of their body weight in vegetable matter. Toward the end of summer, they get real serious, chowing down at every opportunity. That layer of fat they need has caused their weight to double since spring, which is bad news if you have a small garden surrounded by woodchucks. They also endear themselves to commercial vegetable growers, consuming lettuce, beans and squash crops.

Mating occurs after hibernation. The animals separate after mating, the female giving birth to an average of four chucklets about one month later. The family breaks up around the fourth of July, and from this point on, they are solitary creatures. Keep this in mind when trapping colonies. One animal per den complex.

The groundhog den complex has more than one path to the surface. There is usually an highly visible hole in the open, along with another hole that is hidden, maybe camouflaged by rocks, stumps or other natural objects. I have seen where groundhogs in developments will dig under a succession of sheds, from one yard to the next. One of the sheds hides the den, while the rest serve as refuge.

The most obvious sign of the groundhog is their digging activity. There may be a distinct branching trail that leads to a feeding area. A faint trail may show where they are going under a fence. You will have to really look for some holes, as they will not be obvious at all.

PROBLEM: Groundhog living under shed or other structure.

The animal can be captured with a cage trap, then relocated or destroyed. Bait the cage with fresh lettuce or cabbage leaves, and whole carrots. Make a gaudy display of produce, and position the trap close to the hole that you know the animal is using. Stale vegetables will not work, the bait must be kept fresh. They may ignore the bait anyway because of an abundance of natural food. If a week passes without any success, try using one of the commercial groundhog lures or baits. Follow the instructions. If baited sets do not produce, try blind setting. The idea is to position the trap where the animal must pass through it to get somewhere. The double door Havahart trap is required for this set. Typical locations to set the trap are where the animal has been going through a fence, exiting from a woodpile, exiting from underneath a shed, or on a path through cover. Check that there is clearance for the doors to close, and remember to position the locking bars. Stabilize the trap so it will not rock or move when an animal enters. If the animal is captured in a cage, release it as far away as practical. The best release site has some kind of natural barrier, such as a river, that will prevent the animal from returning. Pick a sit where it will not become a problem for someone else. The best alternative is to kill the groundhog.

If you have no success with the cage trap, there are other options. If the underground burrows are away from buildings, gas cartridges can be used to asphyxiate the animal. There are several commercial products sold which release carbon monoxide when lit. For this to work, all of the entrances to the den must be found and covered with boards and dirt, except the main entrance. Light the cartridge and put it in the main hole, then cover it. If you see a groundhog bolting from a hole you neglected to cover, the effort is wasted. I always have my shotgun ready in case this happens (if its OK to shoot). I charge $20 per den on multiple den jobs.

My most effective trapping is done with the 160 or 220 size conibear trap, which kills the groundhog. Some customers will not want this method used for a variety of reasons. I explain to them the different techniques I have to solve the problem, tell them the plusses, minuses and cost for each, then let them decide.

In general, position the conibear trap over the groundhog hole so the animal must pass through the jaws as it departs. If the trap is too close to a structure, the jaws may not be able to close all the way. This distance is about three inches for the 220 size trap. If you have any doubts about the setup, test fire the trap with a stick.

When the animal is removed, secure the shed from further critter use. Refer to GENERAL PROCEDURE FOR INSTALLING AN EXCLUSION BARRIER in the skunk chapter.

PROBLEM: Groundhogs raiding garden.

The effort and expense of keeping groundhogs from eating a garden may exceed the vegetables' worth. Try not to get into trading your services for vegetables. The easiest, cheapest, and most effective solution is to shoot the groundhog, or pay a bounty to someone willing to hunt the culprits. If shooting is not an option, use the trapping methods described in the last problem. A more aggressive strategy is to flood a neighborhood with your advertising, offering a low cost, humane, groundhog removal campaign. Multiple customers in the same location can be profitable in many ways. You will get to be known there as the go to guy (or gal) for all critter problems.

Exclusion is the best way to protect the garden from any potential vegetable thief. Fence climbers can be discouraged with a layer of black plastic secured for a couple of feet on either side of the fence. A strand of electric wire on top of chicken wire will definitely discourage any climbing raiders.

Chapter Six

Tree Squirrels

There are many species of tree squirrels, and all of them are good at climbing, fast turns, acrobatic leaps, tightrope walking on electrical wires and comedy. They instinctively seek out new places to store nuts and build nests. This tendency leads them into attics, chimneys and other problem areas. Most squirrels are beneficial because many of the buried nuts end up sprouting into trees. I believe that squirrels do not remember where individual nuts are buried. Perhaps they look for subtle clues that mark any location, then dig up nuts buried there by other squirrels. This pilfering seems to be accepted practice in squirrel society.

Squirrels provide food for the community of predators. Hawks, owls, snakes, fox, raccoon, weasel, mink, bobcat, feral cat and man all like to eat squirrel meat. Squirrels vocalize with chatter, teeth clicking and barking. When an intruder is spotted, the squirrel becomes all frantic and starts flicking its' tail, barking and scooting around the tree.

Acorns, beech nuts and other tree fruits are favorite foods. Squirrels eat corn by chewing off just the tips of the kernels. They eat most other cultivated crops, mushrooms, pine cone seeds, tree bark and will sometimes eat the flesh of insects and baby birds. They are most active around daybreak and sunset, with the flying squirrel active at night. Flying squirrels eat more baby birds than other tree squirrel species.

Mother squirrel bears two litters, one in March and the other in late July. There are between two and five babies in each litter. Three months after birth, the young squirrels are abandoned by their mother, who gets ready to bear again. Father squirrel only functions as an impregnator, and has no fatherly contact with the mother or offspring.

Hollow places in dead or living trees are favored as nest sites. A woodpecker cavity in a live tree will grow shut over time, so the squirrel chews around the edges to maintain the opening. Squirrels also build leaf nests. These round familiar structures are usually built in the upper branches of trees. Nests will also be built in attics, chimneys and bird houses. Squirrels will go underground in holes, rockpiles and pipes to evade an enemy.

Shells from nuts, pine seeds and other tree fruits in a chimney or attic, are signs that a squirrel has been feeding, and can get in and out. Wood may show sign of being chewed on at the location where the squirrel is entering the house. There may be a nest made in the chimney or attic, and it may contain nuts and seeds. Fiberglass insulation is used as nest material in attics. The clamor caused by squirrels moving around can be considerable. Nighttime noise is an indication that flying squirrels are present, while gray squirrels are active between dawn and dusk.

PROBLEM: Squirrel stuck in fireplace or woodstove.

The uncapped chimney opening is an inviting temptation to curious squirrels. Maybe they see it as just another hollow tree cavity to explore. They end up at the bottom, and jump around from the firebox, to the damper, to the elbow. The squirrel may hide under wood in the firebox, or out of sight behind a joint or curve in the pipe. The only way out is through the living quarters. If there is a direct path from the fireplace/ woodstove, to an outside door, there is an easy solution. First, remove or secure anything that could get damaged (e.g. glassware or knick knacks on shelf). Then, open the outside door, clear a path to it, and let the squirrel find it's way out of the house. If you can see the squirrel in the firebox, close the damper, step aside, and open the fireplace/ woodstove door. Hold a fishing net for protection. If the squirrel is not visible in the firebox, open the damper then open the fireplace/ woodstove door. Step aside and wait. It may take some time for the animal to emerge. Make the room as dark as possible so that the open door is a beacon to the frightened rodent.

If the animal can't be confined to one room it must be captured. This is a more difficult and risky process. The basic tactic is to capture the squirrel, using a net that has a hoop big enough to cover the fireplace/ woodstove door, and a mesh no larger than one-half square inch. If there is no wood in the firebox, the squirrel is visible, and the damper is closed, the job is fairly easy. Wear thick leather gloves. Open the door, and get the net into position fast! Then poke at the squirrel through the net with a stick or wire. As soon as it enters the net, twist the hoop down to the floor, bunch the net around the squirrel and give it a twirl, as you would to a trash bag before putting on the wire twist.

The job is more difficult if there is wood in the box. Keep the net in position as each piece of wood is carefully removed. Get somebody to help (brave souls are hard to find) if the logs are heavy and unmanageable. The squirrel may try to leap out at any time so be ready with the net. If the squirrel should get loose in the room, use the net to either capture it, or shoo it to the nearest open door.

If the squirrel is above the damper, and refuses to come down, tie a rope around the top of the chimney, then lower the other end to the bottom. Eventually, the squirrel may use the rope to climb out. It can be hurried along by using the ammonia trick as described previously under "raccoon in the chimney".

If the rope tactic doesn't work, disconnect the woodstove pipe where it passes through the wall. Then, move the stove out away from the wall, and position the net over the opening as before. The squirrel may be hiding where you can't see it and you must force it to bolt into the net. Push a length of eleven gauge wire through the net towards the back where the stove pipe starts to bend up. Twist the wire around the until the squirrel is pestered enough to bolt into the net.

Cap the chimney before the squirrel is released. If the animal has been burned, injured, severely weakened from the ordeal, or is sick looking, it should be humanely killed.

PROBLEM: Squirrel getting in and out of attic.

It can be difficult to find where a squirrel is getting into the attic. It will fit through a tennis ball size hole, but a smaller opening will be nibbled until big enough. Flying squirrels and chipmunks can fit through very small holes. Once inside they tear up fiberglass insulation, store nuts in various places, and may chew on anything including electrical wires. Some places to look for a point of entry:

- where a chimney joins the roof.
- soffits and fascia.
- louvered vents.
- attic fan openings.
- on the eave line, where a stone wall joins the roof.

What do you look for? Openings in wood joints that look chewed, wood that has dry rot, or is weak from water damage, etc. Go into the attic during daylight hours and turn all lights out. Look for places where sunlight is leaking into the attic. These leaks can be invisibly sealed with blow foam insulation. If the opening resulted from water damage, repair the leak, and replace all weak wood. Screen any vents with a sturdy galvanized quarter inch mesh screen, stapled every six inches into the wood frame around the entire perimeter.

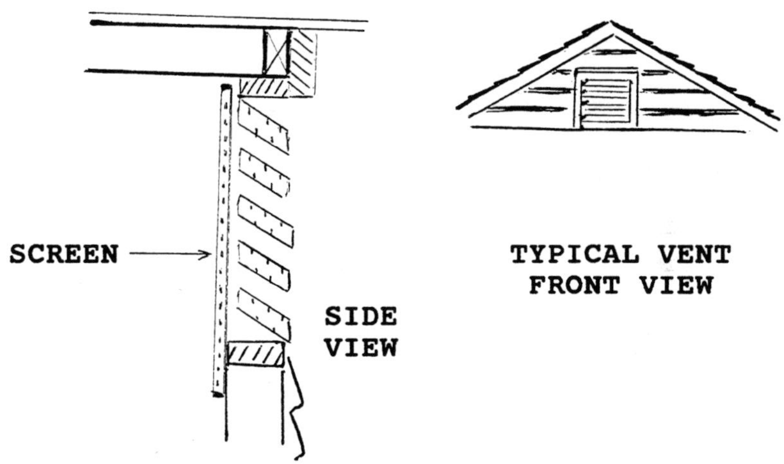

Vent Screen

After every possible opening is secured, there is still the chance that a squirrel is trapped inside the attic with no way out. It will make noise scurrying around, looking for places to chew it's way out. Eventually the squirrel will be hungry enough to enter a Havahart trap baited with peanuts or sunflower seeds. Flying squirrels will readily be caught in the snap type rat trap baited with peanut butter or commercial squirrel bait (has a strong, lasting nut odor). A final concern is to eliminate access to the roof. To get on a roof, squirrels will leap from overhanging branches, walk on electrical wires, and will easily climb on most surfaces. Trim tree branches that extend over the roof, and install barriers on wires that squirrels use: split a two foot piece of thin PVC plastic pipe lengthwise and slip it on the wire.

Squirrels use the same path repeatedly, wearing a trail of scratch marks into the surface over time. Build a barrier at the base of this path, or discourage them by applying a repellent such as ROPEL.

Chapter Seven

Muskrat

Muskrats are common aquatic rodents that benefit wetlands in a variety of ways. They are superb swimmers, good diggers and their reproduction rate is equivalent to mice and rats. Predators relish muskrat flesh, and catch them easily when the muskrat is out of water. Muskrat populations are held in check by predators, including man, along with a variety of diseases. When this balance is altered, muskrat numbers can balloon rapidly causing problems within the ecosystem. Man's harvest of muskrats for pelt and meat is in proportion to market value. When the animal is de-valued, populations can swell because trappers have no incentive to harvest. The high number of muskrats stresses and degrades the available food supply, which has a ripple effect throughout the watershed.

Muskrats eats vegetation that grows in water, such as cattail and bulrush. They will also eat field crops and the succulent plants that grow around waterways. They can cause economic damage to sugarcane and rice crops. They will dig up flower bulbs that are planted near waterways that the muskrat inhabits, especially around ornamental pools and ponds.

The reported averages for litters is two to three per year, with from two to fifteen young per litter. The greatest numbers of young are born in April and May. Muskrats dig into banks to create dens, or make cone shaped huts from vegetation that resembles a miniature beaver house.

Problem: Muskrats Around Swimming Pools

Swimming pools are like any other body of water to some demented muskrats. With the right set of conditions, they will live around pools, doing considerable damage to pool liners, ornamental plants, and digging under decks, skirts and sheds. The only solution is to trap them, usually on dry land with 110 conibears. These traps are small enough that pets and children will not present a problem. The problem may be in conditioning the customer to do some of the work. They can check traps and call you to remove a catch. I had a customer who was so caught up in the problem, he never called me, but removed the catches and reset the traps himself, and still paid what I had quoted.

Problem areas will have distinct trails, and these should be set where the trap does not stick out like a sore thumb. Muskrats seem to be more cautious on land, so the outline of the trap should be broken up using what is convenient and natural.

Muskrat problems can take a long time to solve, don't get discouraged if there is much sign and few catches. Use the time to figure out how the problem can be prevented in the future. Maybe the animals are coming from a nearby waterway, which can be made less attractive using stone rip-rap. Or, the pool area can be secured with exclusion barrier techniques. When you figure it out the customer is relieved, and happy for a solution. Always remember that you were called because pool damages are very expensive to repair.

Problem: Muskrat digging damaging earth dam.

Earth dams and the surrounding banks can be weakened to where the water level is reduced. The only permanent solution is to reinforce the area with rip-rap after trapping as many of the rodents as possible. The basic procedure is the same, but local adjustments may be called for. A galvanized wire mesh catcher is fastened to re-rod anchors around the perimeter of the effected area. This big wire basket is then filled with large stone. The situation may call for lowering the water to do the work, and cementing the structure to some degree. This kind of job takes skills that you may not have, and you may need help in quoting both materials and labor. The job must be worth the effort.

Chapter Eight

Feral Cat

Feral domestic cats cause problems when they overpopulate trailer parks, apartment complexes, junkyards, or any place where food and shelter are available. Females can breed as many as three times a year, with up to ten kittens in a litter. Populations get out of control quickly. It then takes time and considerable effort to get the population under control. In one trailer park I caught 45 cats over a span of two years. Most were caught in winter. I showed the park manager how to operate and bait the traps and he called me when there was a catch. The single door raccoon style cage trap worked better than the double door Havahart, with cooked poultry being the most effective bait, along with commercial cat lure.

There were some problems with this job. Some of the park residents were cat lovers, and generally made the job take longer than it should have. They put out food, closed traps, released cats and complained to authorities. I got assurance from the local JP that these animals had to be controlled with whatever means necessary. There could be unpleasant consequences of having a large number of non-vaccinated cats in the middle of a rabies epidemic, along with a number of other cat diseases such as ringworm, toxoplasmosis and leptospirosis. However, the reason I was initially contacted was not because of the disease threat, but was due to the furor caused by the cats jumping on parked cars and scratching the paint. Most of the residents ignored the cats until their cars were damaged.

On two occasions I have had to destroy barn cats to be tested for rabies. A nursing mother and kittens were attacked by a raccoon. The county Dept. of Health forced the owner to destroy every non-vaccinated cat on the premises. There was a thirty day grace period after which the state would have come in to do the job, but with fines in the thousands of dollars. I used a combination of trapping and shooting to get the job done in three days.

A similar situation developed when a girl was bitten by a barn cat. Her physician gave her a choice of getting the rabies treatment, or having the animal tested. She was not able to identify which of two cats were responsible, so I destroyed both and had them tested. I caught one with a 160 conibear, and shot the other. The job took four hours total.

Feral cats can get suspicious and may disappear for awhile if they feel threatened. Deal with situations as necessary. If an animal needs to be captured within a time period, study the situation thoroughly before attempting a capture. Look for a foolproof way to let the animals habits work for you, such as regular feeding situations and locations where it can be cornered, or funneled into a trap. Always respect that cats are loved by many people and take care to avoid situations that would test your ethics.

Chapter Nine

Bird Problems—Pigeon

Fancy pigeons are raised to compete in shows, and there are races for the homing breeds. Some birds are raised for food. Then there are those who consider pigeons as nothing more than flying rats, because of the problems associated with their droppings. Not only do they look bad and stink, there is also a chemical reaction that occurs with some building materials that leads to rapid deterioration of the building. The feces contain disease organisms that can infect humans, plus there are all kinds of bugs living on pigeons that bite people.

Pigeons can never be totally eliminated since birds from surrounding areas will move in to a vacancy. The only solution I have found is exclusion. The methods I have used to exclude pigeon roosting are porcupine wire and netting. There is a sticky repellent available that I have not had occasion to use, but it requires re-applying periodically, and some weather conditions make it useless.

Netting works better than porcupine wire in some applications. I used netting on a roof next to a dormer pigeons were roosting on. I first secured chicken ladders to the roof where I was going to work.

This made me feel safe, and protected the shingles. Next, I removed all of the loose pigeon crud, feathers and dirt with a brush, then thoroughly scrubbed the roost with soapy water containing bactericide and enzyme cleanser to remove the pigeon odor. This odor attracts pigeons, and will cause them to linger around the screened area.

Measure the area to be enclosed, and cut weatherproof furring strips to frame these dimensions. Wrap netting in the top furring strips and secure with staples, allowing extra netting on the end for seams. Fasten the strips to the underside of the dormer with screws or nails. Next, roll bottom furring strips in the netting until the net is tight. Secure the furring strips to the roof using appropriate fasteners, and seal any fastener holes with roof cement.

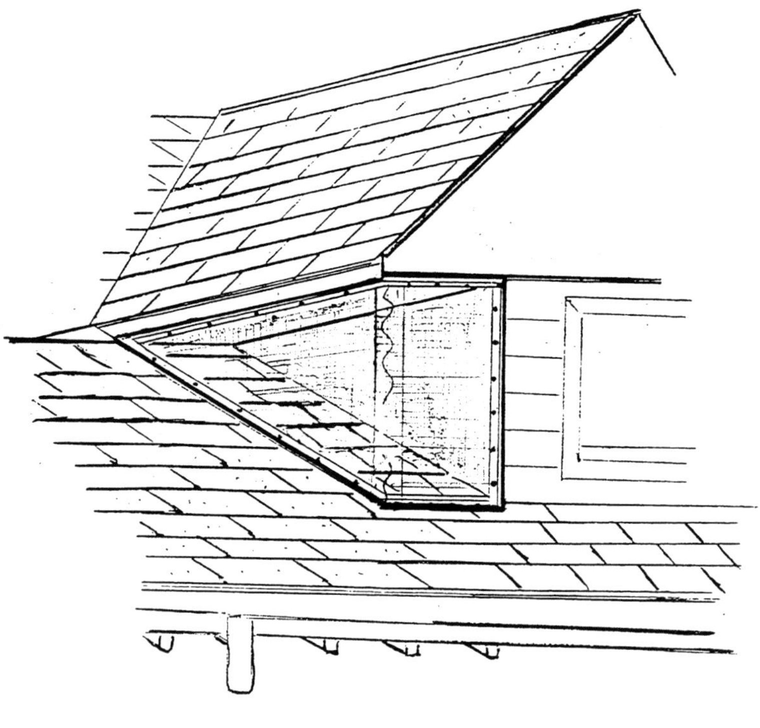

Net Installation Dormer

Each porcupine wire installation is different, the variables being the mounting surface, surface area, location, and the manufacturers specifications for installation. Like most WPC jobs, creativity and improvisation may be required. The manufacturers will provide advice and recommendations, and even work mechanicals for difficult jobs. See the appendix for a list of porcupine wire manufacturers.

Trapping pigeons takes a dedicated effort, and unless exclusion work is done, new pigeons will move in. The key to trapping pigeons is to get the birds used to eating in a safe place where there is always plenty of food available. Introduce the trap after the birds are used to feeding regularly at the station. Set the trap after they become used to it being there, baited with corn inside and out. Some birds may be left inside as decoys. Pigeon trap manufacturers are listed in the appendix. A simple trap can be made with chicken wire on a wood frame.

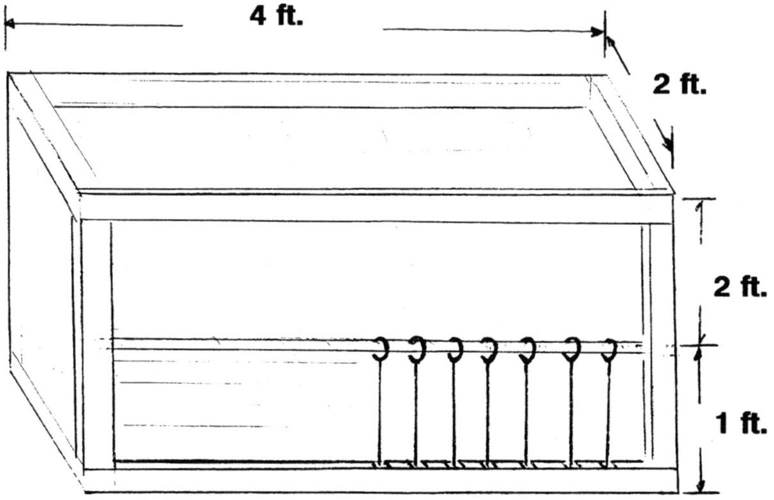

Homemade Pigeon Trap

Bird Problems—Canada Goose

The goose situation here in southeastern Pennsylvania is typical of how, within a few generations, a species can adapt to changing conditions. Our resident population of Canadian geese differ from their ancestors in two obvious ways. They do not migrate, and they spends less time around water. These modern birds live unmolested by predators and hunters. They find sanctuary in urban parks, industrial complexes, golf courses, athletic fields, and other vast expanses of lawn, where they graze on delicious turf. A few hours later they drop a slimy green substance on the lawn which human beings use as something to step in. How are geese discouraged from places they should not be. Here are some elements to consider when dealing with goose problems:

1. Federal protection for, but no permit required to scare, discourage or harass geese.
2. Geese will change their habits to keep using an established food source.
3. Most people love geese, except for those harmed by physical (e.g. airplane collision) or economic (e.g. crop) damage.
4. Geese are big, strong and aggressive. They breed close to water once a year in the spring.
5. Their favorite foods are grains and sprouts.
6. Geese cannot fly during the summer molting.
7. They are most active around sunrise and sunset.

I know of one successful case of goose exclusion. A gentleman farmer had geese defecating all around a small spring fed pond. A large manicured lawn surrounded the pond. Harassing, scaring and hunting in season were tried, but the birds always returned after a short absence. The technique that worked best took advantage of the birds nesting habits. They make a nest near water, where the goslings are led for protection and swimming lessons. The parents will not build a nest if a path to the water is blocked. An impediment was constructed around the perimeter of the pond. Chicken wire on metal stakes circled the pond, one foot from the shore. This was put up before nesting season, and the geese abandoned the pond to nest elsewhere. The fence must have no holes. Geese still fly in and out, but just to eat.

Wealth and fame will be certain to anyone who devises a permanent working solution to the goose problem.

Chapter Ten

Reptiles And Amphibians—Snakes

I make sure the customer understands some basic facts about reptiles and amphibians facts before I agree to do any work. People that believe they have a snake problems often have snake phobias. The phobia is exposed by the way the situation is described, with phrases like, "I saw that filthy creature on the basement floor". People moving to the country seem amazed to find snakes around out buildings, under slabs, in rockpiles, and basements. Old stone houses in particular offer shelter opportunities for snakes.

That the majority of snakes are harmless does not matter to people who loathe snakes. One woman threatened to leave her house and family when she found the shed skin of a large black snake on the dirt basement floor. She could not function in that house until the snake was caught. I never caught the snake, but did manage to convince her that it left the basement through holes in the stone. She paid me to do some exclusion work on the stone walls that I guaranteed would keep the varmints outside. Look for small openings that can be filled with mesh wire, blown in foam insulation, cement or whatever is appropriate. Also check where pipes pass through the wall. Another selling point is that these techniques will effectively exclude rodents as well.

I believe the aforementioned blacksnake shed it's skin then left the basement. I didn't find any sign in the dirt floor, but to appease the customer, funnel boards (1x10's) were placed to guide the snake across rat sized sticky glue traps. All I caught in two weeks was lint and dust. The glue traps are strung together on a board to cover the area the snake is forced to pass over (one foot should do in most cases) and will hold any size snake that gets stuck.

There are situations where snake dens should be removed. For example, children's camps and snake dens don't belong together. Remove the snakes habitat, and the problem is solved. Typical habitat is a rock pile. The rocks have to be moved to get to the den, and the snakes can be caught, sacked and relocated someplace out of harms way. If the rock pile can't be moved, the nest can be burned out. Be sure this can be done safely. Have a helper standing by. Soak the rock pile thoroughly with kerosene. When the fire is out, fill as many of the cracks and crevices as possible with small stone.

There is a commercial snake repellent that claims to repel all snakes for up to two months per application, in any weather. I have not used it, but have been told by a cabin owner in Pennsylvania rattlesnake country, that rattlesnakes are now gone from around the cabin since the repellent application. See the appendix for product availability.

Reptiles And Amphibians—Snapping Turtles

I get a few calls every year to remove snapping turtles from ponds. Two methods get the job done: set lines and traps. The set lines are staked to the bank and baited with fresh sunfish heads. If there are small turtles and water snakes, expect to go through a lot of bait. Use a small hook and you may even catch some of the bait stealers. I use 10/0 long shank shark hooks, filed to a sharp triangular point. Check the point every time you change bait, and change bait every time it gets stinky. It may take some time to reduce the turtle population. In a small one acre pond I had twelve set lines out. There were zero results the first week, then three turtles were caught in one day. That's how it goes sometimes! For this job I charged $35 and kept about 50 pounds of meat. Checking the lines takes time, so in some situations traps would be better. The advantage with trapping is that it can be checked by the customer who calls you when there is a catch.

The first trap I made using weldwire for the body, and chicken wire for the flexible throat.

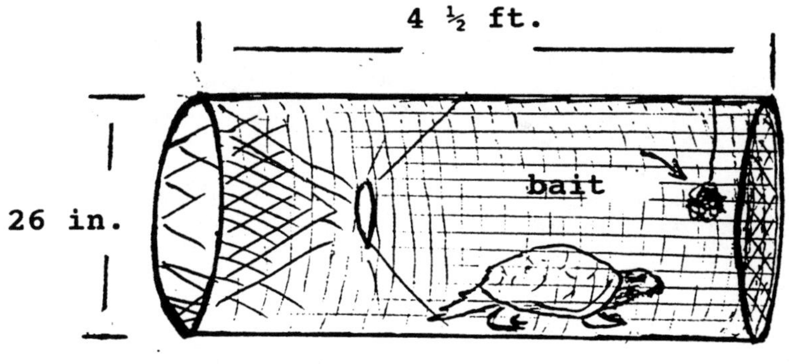

Turtle Trap

This worked until the chicken wire rusted slightly, and a turtle bulled its' way out. Chicken wire will not last in water. I replaced it with poly-net material. Fill the bait cage with fresh fish or chicken. There are commercial hoop traps made entirely from netting. See the appendix for sources.

Handle snapping turtles with caution. I hold them by the tail, and keep the head well away from my body lowering them into a feed sack. I like turtle meat and butcher all I catch. If you want to try it there are several books on catching, butchering and marketing snapping turtles.

Holding Snapping Turtle

Odds And Ends

A Beaver Strategy

I don't have any personal experience with beaver problems, but would like to pass along an ingenious and effective strategy used in Canada that worked to limit flooding caused by beaver dams. The beavers were allowed to stay, because the water level in the dam was fixed by a relief pipe running through the bottom of the dam. The beavers could not plug the pipe because the upstream end was protected by a three foot triangular wire cage. This end was the inflow of the system. The end at the dam was put at the bottom of a break made in the dam. The pipe opening was plugged until the beavers completed their repairs to the break. Then the plug was removed, and the beavers didn't have a clue that there was a hole in the dam. The flexible plastic pipe used is ten inches in diameter and sixty feet long. The three foot triangular cage material is plastic coated weld wire with a one inch mesh.

Bats

You will get a few calls to capture a bat who has lost its' way, and is flying around inside the house causing mayhem and terror.

It is easy to catch and release the creature using a fine mesh net. You can drum up some more business if the chimney needs capping, or you can find some potential entry holes around the eaves, vents, etc. The bat phobic customer will want to avoid future problems by having you seal up all possible entrances. If a colony of bats is residing in the chimney or attic, they will have to be excluded before sealing their entrance(s). Polypropylene netting is suspended over the entrances, which allows the bats to leave, but not return. This should not be done until late August when all of the young bats can fly. The smallest bats can squeeze through a inch diameter hole, so choose the net mesh to match the size of the bat. If in doubt, go small.

Seal up all possible entrances starting with the chimney. Cap it, check the masonry for crumbled spots, and check the flashing for small openings. Next, examine all eaves, vents, pipes, windows and doors. Look for rotten wood, loose flashing, openings caused by wear or sloppy fit, etc. Seal any holes with caulk, wire mesh, blow-in insulation, or whatever is suitable.

Physical Concerns

The joke goes " I treat my body like a temple,... the temple of doom". If your body is used mostly to transport your fat from one eating station to the next, perhaps the WPCO life is not for you. The physical demands can be considerable if you get to climbing chimneys, roofs, attics, and crawling into tight places. While climbing and crawling, you will be lugging heavy equipment, and when you get to the scene of the problem there to greet you is an enraged bull chipmunk or some such creature. You don't want strength or stamina giving out while perched on a high pitch roof. It can really mess up your day if you have an accident or heart attack and nobody is around to help.

If you can't meet the physical demands, there are other ways to be involved in this business. Some ideas: sell supplies, invent tools and things for WPCO's, become a clearing house for WPC information, start a WPC computer bulletin board service. If you want it, your imagination will provide a way.

My Tool Crib

Here is my inventory of tools & supplies:

12 gauge shotgun, 26" barrel, imp.cyl. choke

Ammo: slugs, OO buckshot, #4 copper coated mag.

#6 hi-brass, #8 lo-brass

22 cal. revolver

Ammo: long rifle, short & CB cap

22 magnum bolt action rifle with illuminated scope

Conibear traps, sizes 110, 160, 220, 280 & 330

Cage traps, small rodent size Havahart double doors, to large raccoon single door

Foot hold traps, from size 0 to 4, coil & long spring

Padded jaw foot hold traps, sizes 1, 1 1/2 & 3

Wooden mouse and rat traps

Glue boards, rat size

Bait: large white marshmallows, peanuts, sunflower seeds

commercial sweet baits, put up fish and meat chunks

Lure: commercial cat, raccoon, groundhog and canine

Ladders: 8' folding, 28' extension & chicken (three 8' sections), various fixtures and stand offs.

Rope: 100 ft. 3/8 inch

Safety: glasses with side shields, goggles, ear plugs, belt and harness, gloves

Five gallon buckets, plastic trash bags, feed sacks, burlap sacks, tarps, roll of 10 mil plastic 10 ft. x 100 ft.

Syringe poles & short jab stick, syringes (12 cc) and luerr lock needles (16 & 20 gauge, 1/2 & 1 inch long) and dimethyl ketone (euthanizing fluid)

Concentrated ammonium hydroxide (caustic) & muffin fan: fumes force chimney dwellers to leave

3 ft. choke pole

2 ft. diameter pole net with 1/2 inch mesh, 3 ft. round throw net, 4 ft. x 2 ft square net, insect net

Tape: electrical, masking and duct

Roll of aluminum foil

Wire, 9 & 14 gauge

hand tools: pliers, wire cutters, tin snips, hammer, hatchet, rock pick, hacksaw, vice grips, measuring tape, assorted screwdrivers, wrenches, drill bits, drills, saws, level, staple gun

flashlights, floodlights.

Appendix

Manufacturers:

Mustang Live Catch Traps F.M. Road 1959 Houston, TX 77034-5425 713-480-3203	cage traps for animals, birds, fish, turtles
Lake Superior Guide Service Route 1, Box 32C Dafter, MI 49724 906-635-9089	computer software for trappers
Rob Erickson's P.O. P.O. Box 5345 Glendale Heights, IL 60139 708-858-4895	skunk handling equipment, videos, bait
Nixalite Box 727 East Moline, IL 61244 800-624-1189	bird exclusion materials, repellents for mammals, birds & reptiles
Perkiomen Valley Enterprise Box 84 Pennsburg, PA 18073 800-724-9468	my company - catalog of supplies for WPCOs, + chimney flush system for removing critters

Netcraft
2800 Tremainsville
Toledo, OH 43613
800-638-2723

turtle traps, net and netting supplies

Reed Joseph International Co.
232 Main St.
Greenville, MI 38701
601-335-5822

Bird scaring cannons, bird scaring electronics

Critter Gitters
4317 Nancy Creek Rd.
Kettle Falls, WA 99141
509-738-2230

cage trap designed for skunk capture

Association Publications

National Urban Wildlife News
National Urban Wildlife Management Association
2801 Benson Mill Rd.
Sparks, MD 21152
313-453-8274

The Probe
National Animal Damage Control Association
c/o Wes Jones
Route 1, Box 37
Shell Lake, WI 54871

NACA News
National Animal Control Association
P.O Box 1600
Indianola, WA 98342
800-828-6474

Commercial Publications:
Animal Damage Control Magazine
P.O. Box 224
Greenville, Pa. 16125
412-588-3492

Fur-Fish-Game
2878 E. Main St.
Columbus, OH 43209
614-231-9585

The Trapper & Predator Caller
700 E. State St.
Iola, WI 54990-0001
715-445-2214

Franchise:
Critter Control
640 Starkweather
Plymouth, MI 48170
313-453-6300

Bibliography

North American Furbearers A Contemporary Reference, ed. Eugene F. Deems, Jr. and Duane Pursley. Maryland, 1983.

Howard, Walter E. Animal Rights Vs. Nature. Univ. of California, 1990.

Murie, Olaus J. Animal Tracks. 2nd ed. Boston, 1974

Rue, Leanard Lee. Sportsman's Guide to Game Animals. New York, 1968.

Prevention and Control of Wildlife Damage, ed. Robert M. Timm, University of Nebraska, 1983

Storm, G.L., 1972, Daytime Retreats and Movements of Skunks on Farmland in Illinois, Journal of Wildlife Management, 36:31-44.

Verts, B.J. 1967, The Biology of the Striped Skunk. Univ. of Illinois Press.

Metch, L.D., J.R. Tester & D.W. Warner, 1966. Fall Daytime Resting Habits of Raccoons As Determined by Telemetry. J. Mammal 47(3):450-466

Order Form

Wildlife Pest Control Handbook

How To Run a Successful Full or Part Time Business Working To Solve Problems Between People and Wildlife

Is this your book? If not, why not order one for yourself? Then you can carry it around and use it whenever you want. Can you think of someone who would love to have this book? Why not get it for them now?

Telephone Orders : 1-800-724-9468 COD only

Postal Orders :

PVE
P.O. Box 84
Pennsburg, Pa. 18073

Make checks payable to PVE.

Qty	Book	Price	In PA 6% Sls. Tax	Shipping	TOTAL
_____	Wildlife Pest Control Handbook	$15.95	_____	$3.00	_____

Please print a complete mailing address below.

Name ____

Address ____

City____

State ZIP ____

Phone Number (______________) ____________ - ________________